WRITING AND COMMUNICATION OF ENGLISH SCIENTIFIC PAPERS FOR ENGINEERING MAJORS

工程类专业英语科技论文写作与交流

陈保国　吴文兵　谢　妮　谭　飞　编著

内容简介

工程类专业英语科技论文的撰写和发表不仅是对新理论、新技术、新材料、新工艺的数据记录和结果描述，更是对新成果的展示和交流，其目的在于促进新成果的应用。

本书是"工程类专业英语科技论文写作与交流"的指导性读物。书中全面介绍了工程类专业英语科技论文的选题、构思、写作技巧及投稿和修改交流的方法。

本书共分为11章，主要内容包括：科技论文撰写的前期准备工作、论文定题、摘要撰写与关键词的选择、引言与研究背景的写作要求及方法、正文的结构安排与内容描述、致谢的表述与参考文献的著录、论文图表的格式与单位规范表达、论文表达与语言编辑、论文投稿的方法与过程、论文评阅及反馈与修改、科技论文交流。各部分内容均结合了实际案例进行了分析和阐述。

本书可作为工程类专业研究生的教材或自学参考书，也可供从事相关科研工作的科技人员参考。

图书在版编目(CIP)数据

工程类专业英语科技论文写作与交流/陈保国等编著．—武汉：中国地质大学出版社，2022.7
ISBN 978-7-5625-5293-2

Ⅰ.①工…
Ⅱ.①陈…
Ⅲ.①工程技术-英语-论文-写作
Ⅳ.①TB

中国版本图书馆 CIP 数据核字(2022)第 099034 号

工程类专业英语科技论文写作与交流	陈保国 吴文兵 谢妮 谭飞	编著
责任编辑：龙昭月	选题策划：郑济飞 龙昭月	责任校对：徐蕾蕾
出版发行：中国地质大学出版社(武汉市洪山区鲁磨路388号)		邮政编码：430074
电　　话：(027)67883511	传　真：(027)67883580	E-mail:cbb@cug.edu.cn
经　　销：全国新华书店		http://cugp.cug.edu.cn
开本：787毫米×1092毫米 1/16	字数：243千字	印张：9.5
版次：2022年7月第1版		印次：2022年7月第1次印刷
印刷：武汉市籍缘印刷厂		
ISBN 978-7-5625-5293-2		定价：39.00元

如有印装质量问题请与印刷厂联系调换

前　言

在这个国际化的时代，得益于网络技术和数字化的发展，各专业领域知识的交流越来越频繁，交叉融合越来越快。越来越多的新成果也能够在尽可能短的时间内为学术界所了解。

工程类专业英语科技论文的撰写和发表不仅是对本领域新理论、新技术、新材料、新工艺的数据记录和结果描述，更是对新成果的展示和交流，其最终目的在于促进新成果的推广及应用。

当今国际交流非常频繁，众多科研人员需要利用科技论文与国际同行专家进行学术交流，英语作为一种国际交流通用语言使得英文科技论文的写作变得尤其重要。众所周知，一篇英文稿件能否被国际性学术期刊录用，主要取决于稿件内容的科学性和创新性，但是，稿件的英语写作水平也起着非常重要的作用。如果一篇英文稿件语法错误众多，表达逻辑混乱，语言描述不符合专业英语写作要求，那么，作者所表达的新思想和新观点就很难被审稿人接受，可能导致作者富有新意的成果得不到认可。

事实上，即使在一些英语国家，科技论文写作的教学或培训也很受重视。基于此，为我国从事工程类专业科研人员提供专业英语科技论文写作、投稿和修改等方面的指导而编写此书是一项非常有意义的工作。

本书结合具体案例全面介绍了工程类专业英语科技论文的选题、构思、写作技巧以及论文投稿和修改交流的方法。全书共11章，由陈保国老师和吴文兵老师担任统稿工作，其中，第1、4、5、6、11章由吴文兵老师编写，第2、3、7章由谢妮老师、吴文兵老师、陈保国老师共同编写，第8、9、10章由陈保国老师和谭飞老师共同编写。

在本书的编写过程中，中国地质大学（武汉）刘浩老师、刘鑫老师、梁荣柱老师、涂福彬老师为本书的编著提供了部分素材，研究生李立辰、张云鹏、官文杰、王立兴、闫腾飞、王程鹏、孟庆达、徐昕等参与了本书的编辑工作。他们为本书的顺利出版付出了辛勤的劳动，在此表示衷心感谢。此外，本书的出版得到了"中国地质大学（武汉）研究生高水平课程建设"项目的资助，在此表示感谢。

限于作者水平，书中难免有不足和疏漏之处，敬请读者批评指正。

编著者
2021 年 12 月

目　录

第 1 章　科技论文撰写的前期准备工作 (1)
1.1　概　述 (1)
1.2　科技论文的特点与选题 (1)
1.3　文献调研与科学问题论证 (6)
1.4　科技论文的构思与结构安排 (9)
1.5　科技论文的作者署名 (13)
1.6　学术道德规范 (16)
本章参考文献 (17)

第 2 章　科技论文题目 (19)
2.1　概　述 (19)
2.2　题目的基本要求 (19)
2.3　题目的确定 (21)
2.4　题目的结构类型 (23)
2.5　实例介绍 (24)
本章参考文献 (24)

第 3 章　摘要与关键词 (26)
3.1　概　述 (26)
3.2　摘要的定义和功能 (26)
3.3　摘要的规范化写作 (27)
3.4　关键词的定义与功能 (31)
3.5　关键词的合理选择与使用 (31)
3.6　实例分析 (34)
本章参考文献 (36)

第 4 章　引言与研究背景 (39)
4.1　概　述 (39)
4.2　引言撰写的基本要求 (39)
4.3　引言的组织与结构 (40)

4.4 引言的写作方法与技巧 …… (41)
4.5 实例分析 …… (43)
本章参考文献 …… (51)

第5章 正文其他部分的基本要求与写作要点 …… (52)
5.1 概 述 …… (52)
5.2 研究方法与内容 …… (52)
5.3 研究结果 …… (55)
5.4 讨 论 …… (58)
5.5 结 论 …… (61)
5.6 正文结构化分析 …… (62)
本章参考文献 …… (63)

第6章 致谢与参考文献 …… (65)
6.1 致谢的必要性与内容 …… (65)
6.2 致谢的写作要点 …… (65)
6.3 参考文献的重要性 …… (66)
6.4 参考文献的选取原则 …… (67)
6.5 文献在正文中的使用 …… (67)
6.6 文献在列表中的编排 …… (70)
本章参考文献 …… (73)

第7章 论文图表 …… (74)
7.1 概 述 …… (74)
7.2 图表的作用与分类 …… (74)
7.3 插图的设计制作 …… (76)
7.4 表格的设计制作 …… (78)
7.5 图表的使用和常见问题 …… (80)
7.6 英文图表题名常用句型 …… (82)
本章参考文献 …… (85)

第8章 论文表达与语言编辑 …… (87)
8.1 时态的运用 …… (87)
8.2 语态的运用 …… (88)
8.3 表达方式和技巧 …… (88)
8.4 如何使用地道的英语表达方式 …… (94)
8.5 实例介绍 …… (98)
本章参考文献 …… (103)

第 9 章　论文投稿 ……………………………………………………………… (105)
9.1　期刊的选择 ……………………………………………………………… (105)
9.2　投稿要求与投稿过程 …………………………………………………… (107)
9.3　实例介绍 ………………………………………………………………… (109)
本章参考文献 …………………………………………………………………… (117)

第 10 章　论文评阅意见的反馈与修改 ………………………………………… (118)
10.1　如何分析评阅意见 …………………………………………………… (118)
10.2　如何针对意见进行论文修改 ………………………………………… (119)
10.3　对评阅意见的答复 …………………………………………………… (120)
10.4　如何与评阅人及编辑进行交流 ……………………………………… (121)
10.5　实例介绍 ……………………………………………………………… (122)
本章参考文献 …………………………………………………………………… (132)

第 11 章　科技论文学术交流 …………………………………………………… (133)
11.1　概　述 ………………………………………………………………… (133)
11.2　科技论文学术交流的重要性与形式 ………………………………… (133)
11.3　学术会议展板交流 …………………………………………………… (135)
11.4　国际学术会议大会发言交流 ………………………………………… (137)
本章参考文献 …………………………………………………………………… (140)

附录 A　中国各种基金项目表达方式 …………………………………………… (142)

第1章　科技论文撰写的前期准备工作

1.1　概　　述

科技论文撰写是人类从事科学技术信息书面存储的社会实践全过程,关乎着科研工作完成后能否与同行进行分享、交流、探讨、论证,也关乎着作者的科研工作能否得到正确的评价与合理的回报。科研人员撰写科技论文就犹如厨师做菜一般,前期准备工作就像厨师选材配菜,对整个科技论文质量起着决定性的作用。因此,科技论文撰写的前期准备工作可以说是这全过程的关键步骤,直接影响论文的撰写进度与最终质量。每位科研人员都应高度重视科技论文撰写前的准备工作。

总的来说,科技论文撰写前的准备工作应该包括:掌握科技论文特点与选题技巧、文献调研与问题论证、论文构思与结构安排及作者署名规则等几个方面。同时,在撰写论文前,以及在论文撰写的全过程中,科研人员必须恪守学术道德规范。

1.2　科技论文的特点与选题

1.2.1　科技论文的特点

科技论文是记录科技发展进步过程的文献,也是科技工作者在基础研究、技术研发及工程实践新发展、新成果的书面报告。科技论文已越来越受到广泛的重视,它不仅是科技研究中的重要环节,也是科技信息产生、存储、交流和普及的主要方式。作为科研过程的总结,科技论文的写作水平直接影响着科研成果的价值,同时也是科研工作者能力和作风的具体反映。因此,撰写科技论文是科技工作者的基本功之一,也是科研不可缺少的组成部分。

科技论文具备以下特点[1]:

(1)科学性。所谓科学性就是要求论文资料翔实、内容先进。科学性是科技论文的生命。如果失去了科学性,不管文笔多么流畅、辞藻多么华丽,论文都将毫无意义,而只是人力和时间的浪费。

A. 资料翔实。论文内容、材料、结果必须是客观存在的事实,能够经得起科学的验证和实践的检验。要对每一个概念、数据等进行准确无误的理解和运用,坚持唯物辩证法的立场,实事求是,保持严肃认真的态度,做到立论客观、论据充分、论证严谨。不能主观臆断,更不能为达到"预期目的"而歪曲事实、伪造数据。

B. 内容先进。要求论文能反映所在领域的最新研究成果,其理论和实践水平能够代表当今国内外科技发展水平,如果失去了这一点,论文也就失去了价值。对于综述类科技论文,其内容也应能对所在领域的研究现状和进展进行科学的阐述,并对未来发展方向提供一定的建议。

(2) 创新性。创新是科技论文的灵魂。能否为促进学术领域、行业发展做出贡献是衡量论文水平高低的根本标准。科技论文非常重要的一点就是要有新创见、新观点。科技论文不同于教科书及文学艺术等个性化表达之类的文章,而是在于学术交流,报道新发现,发表新方法、新理论。因此,其内容必须突出"新"字,对于已为人知的观点不必复述,而应突出阐明自己新的观点。

(3) 理论性。科技论文不仅是科学研究的总结,而且是一个再创造的过程。论文不同于一般的科研记录或实验报告,而应提炼出指导科研活动及生产实践的经验教训,发现规律,将之上升为理论,并反过来指导实践。

(4) 简洁性。科技论文要求简洁,不同于一般的文学作品需要各种修辞手段和华丽的辞藻;论文要求行文严谨,重点突出,文字语言规范、简明,能用一个字表达清楚的就不用两个字,不滥用同义词和罕见词。文章应尽可能简短,材料方法部分应简明扼要,结果部分可用较少的图表说明较多的问题,讨论部分不赘述已公认的东西。总之,用最短的文字说明要阐述的问题,以减少阅读时间,使读者用较短的时间获得更多的信息。

(5) 逻辑性。论文的逻辑性是指论题、论点、论据、论证之间的联系一环扣一环,循序撰写,首尾呼应,顺理成章,并做到资料完整、设计合理,避免牵强附会、虎头蛇尾、空洞无物。

(6) 可读性。撰写论文的目的就是进行学术交流,最终是拿来与人分享的。因此,论文必须具有可读性,即语句通顺,结构清晰,所用词汇具有专业性,而且是最易懂、最具表达力的词汇。这样,才能让读者用较少的精力和时间理解论文的观点、结论,并给读者留下深刻的印象。

值得指出的是,除了上述6个方面的特征外,工程类专业科技论文通常还包括针对性和应用性的特征,即针对某个具体工程问题展开研究并进行文献报道,以解决相关工程问题、推广相关工程经验为论文撰写目的,且相关成果可供类似工程推广应用。

1.2.2 科技论文的选题

科技论文是人类从事科学技术信息的书面存储。科技论文的选题一般来源于作者的科研选题,是其整个科研工作的战略起点,是论文成败的关键。科技论文的写作主要涉及两个问题:一个是写什么,另一个是怎么写。选题就是解决怎么写的问题。爱因斯坦曾经说过:提出一个问题往往比解决一个问题更重要。这是因为解决问题也许仅是一个数学上或实验上的技能而已,而提出新的问题、新的可能性,从新的角度去看旧的问题,却需要有创造性的想象力。可以说,科技论文的价值并不在于写作技巧,而在于研究工作的本身,在于选择什么样的课题,取得了哪些有价值的研究成果。如果就论文发表产生的效果和作用而言,只有选择的课题有学术理论和实践应用价值,才会取得良好的科学效果;反之,如果研究论题无必要,即使心血和精力耗费得再多,文字表述得再精彩,也是没有价值的。

1. 选题的作用

总的来说,选题具有以下作用[2,3]:

(1)科研工作的先导。一项科研课题的完成通常有四个阶段:选题阶段、研究阶段、撰写论文阶段、结项验收阶段。其中选题是科学研究的起点,是非常关键的第一步,是整个科研工作的主导思想。论文是以整个科研成果或其中的一部分作为撰写对象的,一个课题完成后,研究者可以撰写一篇论文,也可以由此课题衍生出数篇论文,但论文的选题均在科研课题的内涵和外延之中。

(2)确立研究方向。科学研究活动是一项目的性很强的工作,从提出问题到解决问题,皆是遵循科学合理的辩证过程,只有清晰、深刻地认识与理解问题,问题解决起来才会相对容易。确定选题后,所有的研究工作都将集中某一点上,围绕论题进行。因此,选准论题,即为整个学术活动确立了明确的研究目标,才称得上是真正意义上的科学研究。

(3)界定研究工作范围。现代科学不断向精、向深发展,导致学科分支越来越多,专业方向也越分越细,研究者不可能在多个学科分散发展。在某一时期内研究工作具有一定的稳定性,有较为确定的研究方向和范畴。明确的选题使得这一研究方向和范畴更为具体化,科研工作者可将有限的精力投入到有限的研究目标之中,提高工作效率。

(4)制约着研究进程。研究进程与选题密切相关。在科学研究工作中,由于所选课题的内容、大小及难易程度各不相同,研究进展情况也难以趋同。小而相对容易的课题可在数周或数月内完成,难度大的课题则需要数年甚至数十年。据悉,英国科学家焦耳对"热功当量测定"课题的研究历时四十余载,花费了大半生的精力才取得成功。

(5)决定研究价值。研究的价值在很大程度上取决于初始的选题。有些人忽视科学发展需要,想当然地选择一些根本无研究必要的课题,所获得的"成果"毫无价值;有的人怕担风险引起争议,对重要的现实课题避而远之,只选择些无关痛痒的论题,在降低风险率的同时,其价值也大打折扣;也有些人不愿意深入实际调查研究,选题严重脱离现实;还有些人急于出成果,在功利思想的驱动下选择价值不大的课题,社会效益无从谈起。

(6)制约着研究成果。选题对研究成果的作用体现在两个方面。其一,它关系到科学研究的成败。正如培根所言:"如果目标本身没有摆对,就不可能把路跑对。"即使是伟大的科学家牛顿,其后三十年的研究是建立在唯心主义基础上的,因而难以再铸辉煌。其二,选题一经确定,其未来的研究方向、范畴也随之界定,研究结果的范围基本明晰。虽然在科学发展史上,许多成果为研究者无意中偶然取得,"无心插柳柳成荫"。然而随着各学科领域理论水平的提高,科学研究的偶然因素已越来越少,超出选题范围的"额外成果"会更为鲜见,这是科学深度发展之必然。此外,随着交叉学科领域的渗透,现代科学研究往往需要多人协力攻关,较大的课题还需要较多的经费资助,而选题的好坏成为该课题能否得到资助的决定因素。具有较高学术理论价值或实际应用前景看好的选题,是寻找资助的"敲门砖",容易获得各级各类基金及物力、人力的资助和立项。

2. 选题遵循的原则和方法

选题基本要求就是要提出一个有科学价值、又适合作者个人能力与客观条件的课题。选题水平的高低,既衡量着研究者科研水平的高低,也标志着其学识水平和预见力的高低。因此,不少研究者曾深有体会:提出问题比解决问题更困难。这提示科研人员要有选题意识,要学会发现问题。总的来说,选题可遵循以下原则和方法[2]。

(1)选择本学科发展亟待解决的问题。各个自然学科领域之中,都有一些亟待解决的课题。有些是关系到国计民生的重大问题,有的是该学科发展中的关键问题,有的是当前迫切需要解决的问题。因此,我们必须坚持为社会主义现代化建设服务的方向,选择这些"卡脖子"的课题。

(2)选择本学科处于前沿位置的课题。凡是科学上的新发现、新发明、新创造,它们都有重大科学价值,必将对科学技术发展起推动作用。因此,选题要敢于创新,体现在:①选择那些在本学科的发展中处于前沿位置的、有重大科学价值的课题;②进行前人尚未做过的实证性研究或踏足本学科中他人尚未涉及的新的研究领域;③继续前人未做出的独创性工作,自己发现别人还未发现或尚未引起别人注意的问题等;④进行方法的突破与创新,提出独创性的研究方法、视角;⑤将某一方法应用于新的研究领域,应用不同的方法论进行交叉学科的研究;⑥对一个老的研究问题提供新证据。

(3)选好研究角度(改变选题组合因素)。随着科学技术的发展,许多问题已逐渐为人们所认识,有些课题正在研究之中,这为选题带来一定的难度。学术领域的探究和认识是没有办法穷尽绝对真理的。若从专业研究的现有水平出发,选择有创造性的课题,其重要方法就是要选择新的研究角度。对同一问题,可以从不同的侧面、不同的角度、用不同的方法加以深化研究。

(4)善于运用科学研究中的移植方法。在科学研究中,把某一学科领域中的新发现、新观点、新概念移用到其他学科领域中,为解决疑难问题提供启发和帮助,取得了新的科学发现和技术发明;或者把某一学科或几个学科系统的理论知识、研究方法和技术手段等,综合移植到其他学科领域之中,从不同层次和不同角度去研究和探索,从而做出新的说明,并创造出科学的研究方法。

(5)借鉴别人的选题升华自己的构思。在前人已有的研究基础上可从深度(纵向)与广度(横向)两个方向进行探索。例如,不同地基处理方法设计理论的比较、不同处理效果及质量检验的比较等均属横向结构选题,而对同一种地基处理方法的施工、设计计算、处理效果检验等不同方面的研究则属于纵向结构选题。在借鉴别人的选题时,通过彼纵我横或彼横我纵的方法,研究者可以从中悟出新的选题。此外,还可采用对立而论的思路。例如,在一种新的地基处理方法被提出来以后,多数选题是介绍其成功经验,而对立的观点是观察它对工程结构的长期影响及环境问题。需要注意的是,确立相反的选题要建立在严谨的科学态度和工作实践基础之上,不能缺少科学依据,盲目选题。

(6)注意学科领域研究空白点。科学的使命在于创造,而科技论文的根本任务在于交流学术上的新发现、新设想、新理论、新成就。具有社会价值和学术价值的选题可能是对空白的填补,对通说的质疑,对前说的补充。由于在空白学科范围从事科研工作是平地而起的,要发现空白可从常用的两种途径去探索：一是立题查新，由情报检索部门提供有关信息；二是查阅文献资料，注意发现科学的空白点。

(7) 在争论的焦点中选题。科学是在认识矛盾和解决矛盾的过程中发展的。由于看问题的角度不同，受各种主客观因素的影响，人们对同一问题持有不同观点的现象比比皆是。如果能在选题过程中抓住这些矛盾的焦点，通过理论分析和实践论证提出自己的观点和意见，这样的文章在刊出后更易受到学术界的关注和重视。

(8) 在反思中选题。在科学研究上，问题是带人走出困境的向导，教训是教育人学会创造的老师。从前人失误、失败中汲取教训，明辨得失，慧眼发现被掩盖或被错误判断了的材料结论，纠正谬误，澄清事理，势必会有所发现与突破。我们应学会在前车之鉴中探求新路，在反思中选题。

(9) 结合地区特色选题。针对不同地区的环境特点、经济及社会发展水平、民族居住地域进行分析，根据区域影响所致的不同问题有所侧重进行选题。如东南沿海地区的地基处理方法主要是针对软黏土而发展起来的，对东北地区则是针对冻土而发展，对西北地区则是针对湿陷性黄土等特殊土而发展的。根据地区特点确定地方性科研主攻方向并进行论文选题在可行性及创新性上具有优势，并且是填补地方亟待解决的应用性课题。

(10) 选择预想获得理想效果的课题。选择那些能发挥本人业务专长和利于展开的课题，或者选择那些比较熟悉或感兴趣的课题。这样，可以发挥个人优势。如果课题大小适中，又选准了突破口，就能获得理想的效果。

3. 选题应注意的问题

为了很好地开展课题研究，在遵循上述原则和方法的同时，还应该注意以下问题[2,4]。

(1) 避免盲目性。选题要有科学性、实用性及针对性，必须适应社会实践的需要，以解决实际工作问题和具有理论价值为出发点。课题是研究者经过充分的思想准备和实践准备而提出的，集中体现了选题的科学思维、理论深度和实践能力，反映了命题者的智慧、经验与技巧。我们在审阅的来稿中发现盲目选题、盲目投稿现象比较突出。如社会科学研究文章投到自然科学技术期刊，科普性文章投到专业学术期刊等，作者劳而无获，收不到满意效果。因此，在论文写作开始前，作者首先应当注意选题实现的可能性、选题的实用价值、选题针对的读者对象，对论文的选题应有严密的构思及慎重的考虑，不可草率动笔。

(2) 避免论题过大。科技论文的论题宜小不宜大，不要选择那些无法驾驭的论题。如"排水固结法的计算理论"，这类选题可以写成一本书，而用一本书的题目来写一篇文章，自然不容易写好。千里之山不能尽奇。选题的内容、大小及难易程度要适合自身专业特点，以保证研究工作顺利进行，提高科研成功的可能性，因此，选题的着眼点要放在自己熟悉且体会较深的问题上，注意扬长避短，做到言此事必知此事。长于技术操作者，选题以实验研究为宜；长于逻辑思维、想象丰富者，选题则以理论研究为宜。要根据自己的知识、经验、能力、水平，选择与之相适应的、能胜任并有浓厚兴趣的课题。同时占有的资料要充分，有一定见解。

(3) 避免选题的陈旧性。所选的题目不能是已为同行所熟知或已普遍开展应用的技术内容。这类问题的出现往往是由于文献阅读不够造成的，作者在写论文前未能纵观大局和掌握学科发展动态。

(4)避免选题的重复性。这是研究者中最常见的问题。如一些研究者看到杂志上发表跟基坑监测相关的研究论文后,认为手中也有这方面资料,便将自己某个项目的基坑监测数据汇总后投寄到编辑部。但因缺乏对科技论文含义的理解,未能掌握选题的要领,内容多为千篇一律的数据罗列。由于缺乏科研工作前期设计,不能科学获取资料,无法对研究的问题进行深入分析和科学阐述,难以提出新的观点和认识。

1.3 文献调研与科学问题论证

1.3.1 文献调研

文献调研围绕研究选题收集相关文献资料。虽然在前期选择课题期间,撰写者已对有关资料和学术动态进行了收集和分析,但是在撰写科技论文之前、期间及后期修改过程中仍要查阅大量有关文献作为对已掌握文献的补充。据统计,国内外多数科学工作者查阅文献的时间约占整个科研工作的三分之一。如果没有这些最新的参考文献,要想使论文新颖且具独创性是不可能的。由此可见,文献调研在整个科研和写作过程中的重要性及必要性。

按信息被加工的程度文献可分为三类[5]:①一次文献,包括图书、学术期刊、会议文献、学位论文、专利文献、政府出版物①、产品样本、科学报告、标准文献、档案等;②二次文献,包括目录、题录、索引和文摘等;③三次文献,包括综述和述评等。其中,学术期刊拥有庞大的写作队伍和读者群,出版周期短,内容新颖,论述深入,发行量大,往往能反映有关学科领域研究的最新动态和最高水平,是科研工作者查阅文献最有效且简便的主要来源。

文献调研的范围:①在内容方面,既要有实际资料,又要有有关理论性资料;②在来源方面,既要收集系统资料,又要收集各种零散资料;③在主次方面,既要收集核心资料,又要收集次要资料;④在性质方面,既要收集正面资料,又要收集反面资料;⑤在新旧方面,既要收集新鲜资料,又要收集历史性的资料;⑥在深度方面,既要收集主要资料,又要收集一般资料;⑦在广度方面,既要收集典型资料,又要收集参考资料。

在进行文献调研时,研究者应该重点关注三类人员或机构的文献:①自己导师和课题组已经发表的论文、专利;②国内外同行中在调研领域排名前五团队和学者的论文、专利;③国内外政府部门或企业出版的行业调研报告,如国家自然科学基金委、科技部等部门发布的各类课题申报指南。在此基础上,根据论文的需要,把与科研课题有密切关系并要引用的资料做好记录。

著录文献的出处、作者、题目、杂志名称、卷、期、页数、年代等缺一不可。不要等到文章写好后,到著录参考文献时才发现缺少项目,又得重新查找,白白浪费时间。通过大量的文献调研,从知名学者、质量较高文章的参考文献,顺藤摸瓜,了解经典文献。仔细研读该领域中前人的研究成果,了解同行工作和该领域的研究前沿,掌握研究方向和发展动态,并掌握

---

① 政府出版物包括行政性文献(如法令、条约、资料统计等)和科技文献(如研究报告、技术政策等)。

从事该领域研究的常规研究手段,找出自己可探索的方向。由于各人的习惯不同,文献调研的方法也不同,主要有以下四种方法[6,7]:

(1)滚雪球法。根据内容需要,查找一两篇文献,最好是新近发表的文献综述,然后根据篇末所附的参考文献目录,找到所需要的文献,再从这些文献篇末的参考文献目录找到更多的文献。如此周而复始,参考文献就会越来越多,像滚雪球般越滚越大。这个方法既简单又实用,大部分人都可以采用。此法较省力,但有局限性,易于漏检和误检。

(2)浏览法。经常性地浏览与研究课题相关的近期杂志,跟踪自身领域的研究动态,学习作者的思路、方法和见解,对课题选择和设计很有益处。

(3)跟踪追索法。科研也是一门艺术,成名学者们都有他们可贵的科研思路,他们的经验和教训都可以启发我们的工作。跟踪追索法主要指尽量收集与自己工作有关的某个作者(或团队)过去和现在的论文。在具有系统性的研究里,作者都会引用自己的文章,以表明该研究具有继承性,这样就可以找到许多作者的有关文章。

(4)咨询法。写信向作者、期刊编辑部进行咨询,可能会得到所需要的文献目录、文献资料或有关这方面的知识。这种方法在一般情况下都会奏效,因为这在一定程度上说明被请求者的工作得到了请求者某种程度上的认可,没有理由拒人以千里之外。

文献检索可以分三步开展:
(1)根据研究课题选择检索工具。包括目录型检索工具、题录型检索工具、文摘型检索工具及索引型检索工具。
(2)确定检索方法。包括直查法、顺查法、倒查法、追溯法及循环法等检索方法。
(3)查阅原始文献。

文献检索程序如图1-1所示。

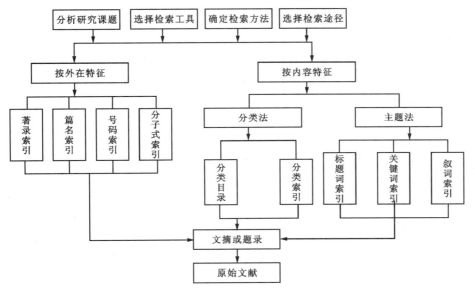

图1-1 文献检索程序图

1.3.2 科学问题的论证

1. 路径

科学问题的论证以科学问题为起点,通过实验、计算、观察、分析、思考、推理和论证等手段获得科学事实,科学地描述原本无法解释的异常现象和客观事实,同时进一步修正、完善和发展科学理论的过程。大量研究表明[8,9],科学问题的提出与论证可以采用"科学问题的界定—科学问题的提出—科学问题的论证—科学问题的论证检验—案例的研读与解析"的路径,并可根据该路径分成相互关联的五个子问题(图1-2),它们在逻辑上层层递进、环环相扣,回答一个完整的核心科学问题。

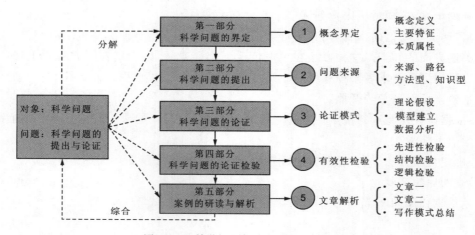

图1-2 科学问题提出与论证的路径

1)科学问题的界定

从学科知识体系或知识树演进的视角清晰地界定科学问题,概括出科学问题的主要特征和本质属性,总结科学问题的判定标准,用简洁明了的公式清楚地表达。科学问题的界定标准可归纳为"unknown+interesting+valuable",即未知的、能引起研究者兴趣的、有价值的问题。

2)科学问题的提出

科学始于问题,科学问题的提出是科学研究"千里之行"的第一步。我国著名地质学家李四光指出:"准确地提出了一个问题,问题就解决了一半。"[8]科学问题提出的重要性不言而喻。那么如何从纷繁复杂的文献和现象中提出科学问题?首先,从科学知识体系内部的逻辑矛盾中探索科学问题的来源,归纳科学问题提出的途径和方法,总结出方法型论文和知识型论文凝练科学问题的通用模式。

3)科学问题的论证

在科学问题提出之后,就要论证科学问题是否成立。这就要归纳出科学问题论证的步骤,总结出科学问题论证的一般通用模式。建议采用总分的体例分别对方法型、知识型文章的过程进行总结和提炼。首先梳理论证的总体思路,其次选取合理的方法,明确研究区域,再进行理论假设和模型建立,最后进行数据分析并得出研究结果。

4) 科学问题的论证检验

在论证科学问题后,需要对它进行先进性、有效性、逻辑等方面的检验。根据科技论文写作的内容,选择正确的检验方式和方法。先进性检验主要从速度、效率和精度等方面进行对比分析;有效性检验主要把研究结果和既定标准进行对比,检验是否在有效范围之内;逻辑检验查看理论假设和概念模型构建的逻辑自洽性。

5) 案例的研读与解析

在了解了科学问题提出与论证的思维过程后,我们还需要通过对经典权威性文章的研读和解析,理解科学问题从提出、论证到检验的整个过程,即分析科学问题是如何提出的、如何论证的和如何检验的。

2. 方法

以上述路径为基础,研究者需要对科学问题提出与论证的过程进行反复思考和研究,可采用以下几种方法[8-11]。

(1) 文献研究法。收集国内外相关研究资料,了解研究动态,为深入研究提供借鉴,启发研究思路,清晰地界定科学问题。

(2) 归纳总结法。对方法型论文和知识型论文中科学问题的凝练方式与思维过程进行归纳总结。

(3) 演绎法。通过研究假设、模型构建、数据分析和研究结果的阐述对科学问题进行论证。

(4) 对比分析法。将研究结果与相关标准或者与以前的研究结果进行对比分析,证明论证的有效性。

(5) 案例分析法。对已发表的高质量论文进行案例解析,加深对提出和论证科学问题的理解。

1.4 科技论文的构思与结构安排

1.4.1 科技论文的构思

科技论文的构思必然有它自己的要求,体现在正、反两个方面。这里,先从它的反面来说,即科技论文在构思上要注意四个"不要"[12]。

(1) 不要把论文写成教材或讲义。有的论文作者当了多年教师,撰写了很多教材或讲义,习惯成自然,一提笔就从 ABC 写起。用这样的方法来写论文,反而容易掩盖其创新性。需要明确的是,科技论文的目的主要是交流新成就,报道新发现、新发明,发表新设想、新理论,提出新定理、新模型,探索新方法、新条件。科技论文贵在一个"新"字,一定要围绕"新"

做文章,以各种方法有效地推出这个"新"字。

(2)不要求大求全,"自成体系"。上面说过,科技论文力求其"新",而不是着力于求大求全,这点是容易理解的。我们经常看到,一心自成体系的文章,往往徒有其表,反而思想虚空。比如大段复述已有知识、重复推导已知数据反而会造成累赘,只需注明参考文献即可。

(3)不要故弄玄虚,"浅入深出"。深入浅出是可贵的,而相反的"浅入深出"容易造成读者的困惑。

(4)不要搞类似翻译的长句子,"洋里洋气",读起来疙疙瘩瘩。不是说不许出现长句子,而是说不要一味地去"搞"不必要的长句子。句子的长短要根据对意思的表述而论,但是我们民族语言的习惯,是求短不求长。

以上四点,前三点是讲全篇的构思,最后一点是讲执笔成文时的思路走向。总的来说,要求读者掌握两个"着眼点":一是着眼于科技论文的读者是专业上的内行,是作者的同行,作者动笔写论文的时候,心目中一定要有这样的读者存在;二是着眼于国际流行的英文科技论文,其读者当然也来自各个国家,在文风上应该力求保持学术英语的特色。

接着,再从正面来说说科技论文在构思上须注意的四个"要"[12,13]。

(1)文章的构成要使人容易理解,在学报的编辑和作者之间应该有这样一个统一的认识:写得难懂,是文章的一大缺点。对于这一点,读者自然是十分赞成的,问题就在于编辑和作者,特别是作者。如果作者舍得放下架子,不求形式上的"高深",不徒慕虚名,这一点还是可以统一认识的。为了便于读者理解和接受,就要讲究文章的构成方式,比如可否考虑这样几点:从已知的到未知的,从简单的到复杂的,从具体的到抽象的,从读者兴趣浓厚的到兴趣较少的,从读者容易赞同的到不大容易赞同的,从历史的到现实的,从当今的分析到未来的预测,等等。当然,对各方面切忌面面俱到,也要具体情况具体对待,必须突出重点,推出新意。

(2)写法要明快简洁。比如,有并列关系的宜分项证明、分条标码;用图表更清楚的,就用图表;各个逻辑段落宜有主旨句居于句首,作为概括;层次要清晰,不宜重床叠架。

(3)标题要具体、明了、概括。不要随便大范围地使用如"……的研究"之类的短语作标题,以免不得要领;也不要用过长的标题,过长的标题可以分解为正标题、副标题;也不要用类似产品名称的名词作标题,比如"……打桩机""……控制仪"等。总之,标题要表明论文的核心或研究主题,起到"画龙点睛"的作用。

(4)引文要讲究质量。至少要做到这样几点:不可断章取义,曲解原著;要把引文和作者的见解严格区分开来;核对无误;不可失密。另外,还建议尽量少引。

1.4.2 科技论文的结构安排

工程类科技论文与一般科技论文相比,除应有一般科技论文的共性外,还应具有:①专业性,工程类科技论文应围绕某一工程领域的相关主题展开论述,具体包括模型试验和现场试验、理论计算、数值模拟等方面的论文;②工程专业的多样性,如土木工程、地质工程、勘查技术与工程等;③研究成果的实用性、真实性,与其他科研成果相比,工程类研究成果更具实用性,通常是为解决实际工程中存在的某些技术难题而进行的科研活动,更加强调科技成果的实用性。

工程类科技论文主体部分的基本构成,与一般科技论文的编排格式基本相同。国家标准《科学技术报告、学位论文和学术论文的编写格式》(GB/T 7713—1987)[14]中指明,报告与论文由前置部分和主体部分两部分组成(图1-3)。以下从综述型、试验研究型、理论研究型和工程实践型探讨工程类科技论文的主要结构[15]。

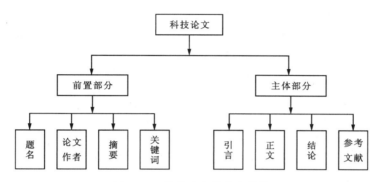

图1-3 科技论文的基本结构

1. 综述型

图1-4为综述型论文主体部分的基本构成。此类论文虽一般不要求在学术上有所创新,但综述型论文不能简单罗列已有成果,应在综合分析和评价现有研究成果基础上,分析总结特定时期某一研究领域的规律与发展趋势,侧重于回顾、评述、分析与展望,提出有根据的、合乎逻辑的、具有启迪性的建议和意见。这种论文的撰写要求较高,应具有该学科专业方向的权威性,能对某一学科方向的发展和研究产生导向作用。

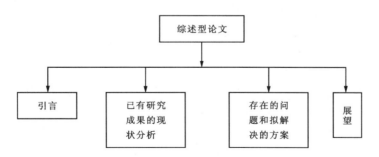

图1-4 综述型论文的基本结构

2. 试验研究型

图1-5为工程类试验研究型学术论文主体部分的基本构成。此类论文大多是为解决某一实际工程问题而进行的科研活动,要求具有创新性、科研成果的前瞻性和实用性。论文内容要求原始数据翔实、论据充分、结果可靠。特别是在论文的试验概况部分,作者应提供

试验设计翔实的原始数据、材料参数、实施方案等,这些信息可为原创成果的真实性、合理性提供重要参考。同时,试验研究应具有再现功能,即按论文所述的基本信息能够再现原试验。

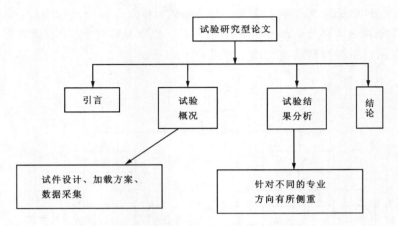

图1-5 试验研究型论文的基本结构

3. 理论研究型

图1-6为理论研究型论文主体部分的基本构成。此类论文侧重于理论推导的结构性、条理性,明确提出理论创新性,体现研究者的原创性、前沿性。为体现理论分析结果的正确性、合理性,通常应有与理论成果相比较的手段,如试验验证或与他人研究成果进行对比分析。

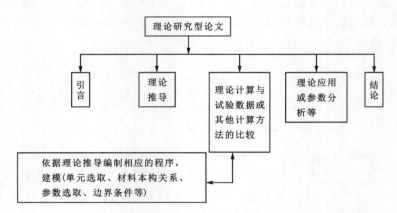

图1-6 理论研究型论文的基本结构

4. 工程实践型

图1-7为工程实践型论文主体部分的基本构成。此类论文研究对象为国内外重大工程,应着重阐述工程的特色实际,如在工程设计中应采用何种先进技术,解决了哪些关键技术等,避免一般性的介绍。

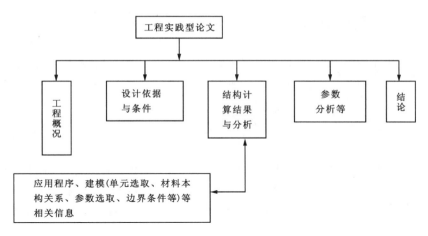

图1-7 工程实践型论文的基本结构

工程实践型论文应从以下几个方面展开论述：工程设计所涉及的设计依据与条件、计算与分析等。目前，国内的重大工程，为确保工程设计安全、经济、合理，通常采用计算机进行辅助模拟和设计。这就涉及应用程序（如 ABAQUS、ANSYS、FLAC、PFC 等）数值模拟软件的选择、模型的建立、边界条件的模拟等诸多问题，这些都应有详尽的说明。某个重大工程的成功完成对类似项目具有借鉴意义，这些细节信息的介绍能够反映该工程设计的科学性、合理性。

1.5 科技论文的作者署名

1.5.1 作者署名的目的

随着科学技术的发展，越来越多的研究课题需要由多领域、特长各异的专业人员共同完成。因此，尽早确定作者名单对收集与整理资料、写作分工等十分重要。此外，合理的作者署名与排序不仅可以反映出各作者的贡献大小，同时也表明哪些人应该对所发表研究成果的科学性和真实性负责。

在科技论文中，署名的目的如下[16]。

(1) 表明作者的劳动成果及作者本人获得社会的承认和尊重，并对该论文拥有著作权。这既是应得的荣誉，也是论文版权归属的声明。

(2) 标明论文的责任人，是文责自负的承诺。论文一经发表，署名者将负有政治上、科学上和法律上的责任，而作者署名的位置前后则表明了他们对论文所报道的结果与论点所负的责任大小和贡献大小。

(3) 便于编辑、读者与作者的联系，沟通信息，互相探讨，共同提高。

1.5.2 作者署名的资格

《中华人民共和国著作权法》规定:"著作权属于作者。"因此,作者对成果有优先权,是论文法定主权人,也是论文的负责者。关于署名的资格,目前国内外的普遍看法如下[17]。

(1)作者应自始至终参与该项研究工作。

(2)作者应能对该项研究成果具有答辩的能力。

(3)作者必须参与研究论文的撰写工作。

(4)作者必须阅读过论文全文并同意发表全文,明确承担由此而带来的各种责任。论文的作者不但要对自己所负责的实验数据的真实性和准确性负责,并且要对该文的结果、结论负责。如果论文中出现了有争议的内容或更重要的问题,每一位作者都必须承担责任。作者署名,责任是第一位的,其次才是荣誉。

特例说明[16]:《美国医学会期刊》(*Journal of American Medical Association*, *JAMA*)要求作者就以下三个方面进行说明:①参与实验,包括在酝酿和设计实验、采集数据、分析和解释数据等方面做了哪些工作;②论文撰写,包括在撰写论文、对论文的知识性内容作批评性审阅等工作中做出了哪些贡献;③做了哪些支持工作,包括统计分析,获取研究经费,行政、技术或材料支持,以及指导、支持性贡献和其他贡献。*Nature*、*Science* 对署名作者也有类似要求,以防止论文署名时出现"友情作者""名誉作者"。

至于为作者争取到资助、提供技术服务、提供样品材料或负责行政管理的人,虽然他们对完成论文不可或缺,但仅这些支持性工作本身不足以使他们成为论文作者。对论文曾经提出过有益意见的人,也不一定列为作者。以上所列举的人员只是间接地与论文的研究工作相关,应在致谢中表明他们的贡献并对他们表示感谢。

1.5.3 作者署名的排序

论文的第一作者一般应是具体工作的主要执行者,有时也可以是整个研究工作的主要设计者,其他作者的署名排序应该按照贡献来确定。作者的排序,也是一个严肃和认真的问题,最好在论文撰写之前,甚至研究工作进行之前或进行的过程中,就通过充分讨论予以确定。注意作者的排序必须得到所有作者的认可。通常第一作者是论文的执笔撰写人,再按照作者对研究工作的贡献大小来排出顺序。

1.5.4 通讯作者

在国外,通讯作者(corresponding author)是所有作者中的一个例外,通常由资深作者(senior author)担任,通常是实际统筹处理投稿和承担答复审稿意见等工作的主导者,也常是稿件所涉及研究工作的项目负责人。通讯作者多位列作者名单的最后并使用符号(一般用"*")标注说明,但其贡献不亚于论文的第一作者。

目前没有对通讯作者的官方定义和解释,但学术界学者普遍认为通讯作者具有两个方面的功能,既是论文的对外联系人,又是论文的负责人,即对论文内容的真实性,数据的可靠性,结论的可信性,是否符合法律规范、学术规范和道德规范等方面负责。只有通讯作者与第一作者不一致的情况下,才需要标注通讯作者。国际惯例是,在多作者署名的论文中,第一作者往往是承担主要实验工作的人,在很多情况下,是某一实验室的研究生;而通讯作者(即资深作者)才是主要学术思想的提出者,并是读者对有关论文提出各种问题时能与之讨论和联系的作者,通常还是课题负责人[18]。

科技论文的第一作者往往是研究中承担大部分实验的研究人员,对该部分实验数据的真实性和准确性负有全责,负责撰写并完成论文初稿并在导师的帮助下进行论文修改。只承担实验而不撰写论文的作者,通常不应排名第一。此时,应由论文的思想提出者兼论文撰写者直接承担第一作者和通讯作者,由于通讯作者与第一作者的贡献和责任合二为一,在这种情况下就不需要另列通讯作者(也可以标注出来)。论文的第一作者虽然贡献很大,但并非论文的最大贡献人和唯一责任人。将第一作者和通讯作者的作用区分开更能明晰第一作者与通讯作者对文章的贡献,其责任和义务的界限也更加清晰明了。在国内,有些单位评奖或登记科研成果时,通讯作者与第一作者往往被同等看待。在科学界,被标注为通讯作者的人往往更能赢得尊敬的目光。很显然,通讯作者这个角色的出现实际是第一作者的角色发生了分异,第一作者应负的责任也部分地转移到通讯作者身上。这也就是说,第一作者的一些责任和义务被通讯作者承担了,但这不妨碍他的论文排名位于在最后[19]。

例:

2006年获得"诺贝尔生理医学奖"的Andrew Fire和Craig Mello分别是第一作者和最后一位作者[16]。诺贝尔奖评审委员会认为,1998年发表在期刊 *Nature* 的论文 *Potent and specific genetic interference by double-stranded RNA in Caenorhabditis elegans* 报道了发现双链RNA引发基因沉默现象,展示了他们突破性的研究进展。在这篇文章中,Andrew Fire是第一作者和通讯作者,而Craig Mello排在最后。因为欧美体系的生物学杂志往往认为senior author排在最后,所以Craig Mello才有可能以末位作者的身份与Andrew Fire共享诺贝尔奖,而其余作者则与诺贝尔奖无缘。

1.5.6 作者单位或通信地址的标署

《中华人民共和国著作权法》规定,科技论文的作者享有署名权,著作权的其他权利归著作权人所在的单位享有。因此,科技论文中的作者必须署有单位名称[20]。

作者地址有助于作者身份的识别,有助于提高作者单位的知名度,同时也为科技论文提供负责单位,也便于期刊编辑和读者与作者联系。在署名作者单位时应注意以下几点。

(1)署名应为单位全称,邮政编码必不可少;高等院校的通信地址需具体到院系和研究室,以避免与论文相关的邮件无法寄达作者本人。

(2)如科技论文的数位作者均来自不同单位,则需对不同的单位署名编号(如 a、b、c

等),作者地址的排序与作者排序一致,并将地址编号以上标的形式标注在相应的作者署名处,一定要清楚地指明作者的有效通信地址。

(3)一般第一作者所在单位位列署名单位的首位,如第一作者或通讯作者同时为其他单位的兼聘或客座研究人员,为体现成果的归属,需要在论文中同时署名实际所在单位和兼聘单位的地址。

(4)有些作者除了在本单位工作之外,可能在其他单位担任兼职教授或研究员,也有可能主持了其他单位的重点实验室基金,根据需要,也可将担任兼职工作的单位或者基金支持单位同时列出。

例[21]:

Title:Three-dimensional consolidation theory of vertical drain based on continuous drainage boundary

Journal:*Jorunal of Civil Engneering and Management*

Authors:Yi Zhang[a], Wenbing Wu[a,b], Guoxiong Mei[b], Longchen Duan[a]

Affiliation:

[a]Faculty of Engineering,China University of Geosciences,Wuhan,430074,China

[b]College of Civil Engineering and Architecture,Guangxi University,Nanning 530004,China.

1.6 学术道德规范

学术腐败有很多种表现形式,例如制造学术泡沫,搞假冒伪劣,抄袭剽窃,钱、学、权的三角交易,从事注水学位教育或培训,在学术评审时拉关系,以及办刊收费,等等。学术规范问题已经由一个单纯道德层面上的问题日益演化为一个跨道德与法律两个层面的重要现实问题。作为高等教育最高层次的培养人才,研究生在形成良好的学术风气、塑造浓厚的学术氛围、建立诚信的学术机制、培养自觉的学术意识及提升高校学术层次等方面肩负着历史的重任,理应严格遵守《中华人民共和国著作权法》《中华人民共和国专利法》《科技工作者道德行为自律规范》等相关法律法规、社会公德及学术道德规范。

在展开科研工作的同时,我们应遵守下述基本学术道德规范[22]。

(1)坚持实事求是的科学精神和严肃认真、一丝不苟的科学态度,严格遵守国内外公认的论文写作规范,诚实,求真,尊重他人劳动成果,遵守国家有关法律法规。

(2)坚持公开、公正、严谨、自律的论文创作过程,防止和杜绝发生抄袭、剽窃、造假等有损个人名誉、学校声誉和学术道德的不良行为或重大学术失误。

(3)科技论文的署名应实事求是,合作研究成果在发表前要经过所有署名人的审阅,并签署确认书,署名作者及所在单位应对该项成果承担相应的学术责任、道义责任和法律责任。杜绝出现"友情作者""名誉作者"及"通讯作者泛滥"等不良学术行为。这类行为严重违背了学术规范、学术伦理和知识产权,是一种署名造假行为,是严重违反学术道德的行为。

(4)科技论文应是作者亲自参与资料收集、选题、构思方案、深入研究、周密思考、精心撰写、反复核查后获得的具有创新性或实用性的知识成果。防止和杜绝粗制滥造、抄袭和一稿多投的不良行为，一经发现，必当受到法律的严惩。

（5）科技论文中引用他人研究成果，包括方法、观点、数据、结论、公式、表格、图件、程序等必须注明原始文献的出处，不使用未经亲自阅读过的二次文献；所有使用过的文献应在文后参考文献中翔实列出，避免遗漏和错误，防止和杜绝侵害他人知识产权。

（6）在对自己或他人的作品进行介绍、评价时，应遵循客观、公正、准确的原则，在充分掌握国内外材料、数据基础上，做出全面分析、评价和论证。

（7）科技论文已有一种文字发表后，如需用第二种文字进行二次发表时，必须注明第一种文字已在何时何种文字版期刊、论文集、网站等媒体上发表。

本章参考文献

[1] 古斯塔维. 科技论文写作快速入门[M]. 李华山,译. 北京:北京大学出版社,2008.
[2] 董琳. 科技论文选题的思路及方法[J]. 中国计划生育学杂志,2007(3):189-191.
[3] 姚福琪,姚咏梅. 科技论文的选题与写作[J]. 保定师范专科学校学报,2002,15(4):68-70.
[4] 丁力. 科技论文选题的原则与方法[J]. 河北电力技术,2007(6):63.
[5] 拉姆奇. 如何查找文献[M]. 廖晓玲,译. 北京:北京大学出版社,2007.
[6] 任银玲,张莉. 中文农业科技文献检索的常用方法和技巧[J]. 河南图书馆学刊,2011,31(4):66-68.
[7] 高丁丁. 浅谈科技文献检索的技巧[J]. 教育观察,2016,5(19):135-136.
[8] 王淑芳. 论文写作中科学问题的提出与论证[J]. 语言文学研究,2017,23:36-37.
[9] 黄冰. 科技论文的主要论证方法[J]. 安徽工学院学报,1988,7(4):144-150.
[10] 姚远. 科技论文的论证分析方法[J]. 曲阜师范大学学报,1989,15(2):107-109.
[11] 徐正林. 课题设计与论文论证的几个基本问题[J]. 写作讲堂,2018,5:106-109.
[12] 杨松浦. 科技学术论文的写作:构思与语言[J]. 学报编辑论丛,1990,111-115.
[13] 王绍汉. 学术论文的构思[J]. 山东轻工业学院学报,1992,6(4):74-80.
[14] 全国信息与文献标准化技术委员会. 科技报告编写规则:GB/T 7713.3—2014[S]. 北京:标准出版社,2014.
[15] 李淑春,张汉平. 建筑结构科技论文主体部分基本构成[J]. 编辑学报,2011,23(S1):90-92.
[16] 杨雪,余磊,王开云. 科技论文不同署名作者的贡献及责任[J]. 昆明学院学报,2011,33(6):113-116.
[17] 张久权. 科技期刊论文写作系列讲座:Ⅲ. 题名和作者[J]. 中国烟草科学,2009,30(5):81-82.
[18] 邹承鲁. 我的科学之路[M]//邹承鲁杂文集. 北京:学苑出版社,2008.

[19] 韩翠娥. 科技期刊应规范作者署名和单位署名[J]. 编辑学报,2016,28(S2):S11-S12.
[20] 任胜利. 英语科技论文撰写与投稿[M]. 2版. 北京:科学出版社,2011.
[21] ZHANG Y,WU W,MEI G,et al. Three-dimensional consolidation theory of vertical drain based on continuous drainage boundary[J]. Journal of Civil Engineering and Management,2019,25(2):145-155.
[22] 戴,盖斯特尔. 如何撰写和发表科技论文[M]. 6版. 北京:北京大学出版社,2007.

第 2 章　科技论文题目

2.1　概　述

在阅读科技论文的时候,读者最先看到的就是论文的题目。他们通常会按照个人需要和兴趣选取部分内容进行阅读。这时,题目起着指示论文内容的作用。研究者在进行文献检索的时候,从文献检索系统的题录或摘要等索引里首先找到的也是论文标题。文献工作者在编辑索引、文摘或题录时,要根据题目所指示的论文主题进行取舍和分类。论文作者在写作之前要选题和命题,写作完成之后要对照论文内容修改题目,使论文内容切题,使题目简短而且准确概括论文的内容[1]。期刊编辑在阅读、编辑科技论文的时候,要按照读者和检索的要求修改论文和题目。

总之,科技论文的题目起着准确概括论文主题和指示论文内容的作用[2]。因此,应该精雕细琢论文题目中的每一个字(词),以及它们之间的顺序,并在字数限制范围内尽可能准确而精简地给出接下来正文的主要内容。

2.2　题目的基本要求

好论文的题目应该是文章内容最简短的表达形式,并能覆盖全部的内容,看了题目就可大体上判断论文要论述哪方面的内容。最好在论文工作进展到一定的深度再敲定题目。题目的各方面作用决定了人们对它的要求大致有以下几个方面。

2.2.1　意义要确切

论文题目要准确地概括论文主题和指示论文内容,使读者从题目能推测出内容。要尽量做到题目内涵和论文内容一致。例如,《软黏土地基中PHC管桩负摩阻力分布规律研究》,从标题看,论文内容涉及负摩阻力问题,而且只涉及软黏土地基中的PHC管桩,没涉及其他桩型。论文主要关注负摩阻力的分布规律,不包括监测和消除负摩阻力的方法。如果论文所述的正是这些内容,则题目与论文内容贴切,即内容切题。

如果把上述内容的题目写成《软黏土地基中桩的负摩阻力研究》或《负摩阻力分布规律研究》,则就该论文的特定内容而言,这两种标题所指内容大得多,因而太空泛且不确切;反之,把题目范围写得过窄也不贴合论文内容。

2.2.2 特点要突出

科技论文的一个显著特点是具有新意,作为它的题目当然要体现其新内容,突出本论文研究内容或方法的特点,避免一般化、雷同化。没有特性的题目不应成为科技论文的标题,提出题目的特性就是突出本论文的新贡献,给读者以新的信息。从这个意义上可以说,没有特性的标题就意味着文章没有新的内容[3]。

一个具有突出特性的题目可以使文章更加醒目和鲜明,更能吸引读者。人们总是要了解新的知识,对含有新意的标题读者不会不发生兴趣,总要看个究竟。例如,《土塞效应对管桩低应变测试视波速的影响研究》[4]这个标题有两个很重要的关键词——土塞效应和测试波速。之前的研究从来没有人注意到土塞效应会对测试波速产生影响,因此这个题目就能很直观地激发读者的思考:为什么土塞效应对测试波速产生影响?是怎么影响的?凡是桩基测试领域的研究人员,看到这样的题目不会一扫而过,必然发生兴趣。然而,作者最开始拟定的题目是《土塞效应对管桩检测信号的影响研究》,而该领域研究人员都知道土塞会对检测信号产生影响,因此,这个题目没有反映出文章的中心内容,使读者看不出文章的特点,显得过于平淡无奇,自然也没法引起读者的注意。

值得注意的是,有的论文是在他人工作的基础上进一步得到的某种研究结果,论文的题目应根据自己的工作拟定,但有的作者拟定的题目却与所列参考文献的题目几乎相同,就失去了特性,有可能导致论文发表失败[3]。

2.2.3 文字要简练

在意义确切和特点突出的前提下,论文题目的文字要十分简练,必须用高度概括的语言表达文章的主题。要使标题简练,至少应注意三点:①非文章主题的内容不要纳入题目;②语句必须精炼,一切不反映实质内容的词都可以去掉,只保留那些代表文章中心内容、体现文章特点的关键词和必要的语法词(包括限定词和虚词);③文字结构要紧凑。

原则上,论文题目越短小精悍越好,一般中文题目字数以在20字以内为好,英文标题一般以15个单词左右为宜[5,6]。例如,一篇工程力学方面的论文[7],原来的题目为《建立完全无约束的三类变量的广义变分原理的泛函的一种方法》,这样的题目一连用了四个"的",显得太冗长了,不妨改为《无约束三类变量广义变分原理的新方法》,一个"新"字就道出了此论文的根本意义。有些题目写得过长,往往是加了"关于……的研究""关于……的调查""关于……的探讨""关于……的报告"等多余的话[8]。例如,《关于软黏土地基中PHC管桩挤土效应问题的分析和探讨》,如果改为《软黏土地基中PHC管桩挤土效应研究》同样能使读者从标题中了解该论文的内容。

当然,论文题目的字数也不能一概而论,还要依据具体情况决定。有时简化题目会减少它所提供的信息,这需要权衡处理。同时,题目过于简化势必造成意义笼统,适用范围过宽,使读者看不出文章的具体内容。因此,简化题目应以不减少它应包含的关键内容为条件,如果简化后的题目反映不出文章的特点和内容,则宁可长一些。在论文题目不能简短的情况下,也可以考虑增加副标题,但一般科研论文除批驳性论文外,尽量不要使用副标题[6]。

2.2.4 结构要合理

科技论文的题目通常为偏正结构或动宾结构。例如:①《海岸带农林复合生态系统建立技术》属于中心词("建立技术")加限定词语("海岸带农林复合生态系统")组成的偏正结构;②《排水固结法减少工后沉降的试验研究》属于主("排水固结法")谓("减少")宾("试验研究")齐全的动宾结构;③《改进单纯形法优化材积方程及其数值解》属于动("优化")宾("材积方程")结构,这是一个少见的主题("优化材积方程")、研究方法("改进单纯形法")和重要结论("数值解")同处在一个题目中的例子[8]。

2.2.5 语言要规范

科技论文是对客观事实、客观规律的描述,必须严谨求实、一丝不苟。这反映在题目上,则要求题目语言表达规范、形式安排考究。语言表达规范是指在拟定题目时不能生搬硬套、盲目猎奇,尤其是要文理通畅、用法严谨、专业术语规范,不能别出心裁地自己杜撰,也不能太过于口语化,要用公认的、稳定的、普遍适用的词[7]。形式安排考究是指标题的结构形式要根据论文的研究对象、内部结构安排和行文目的等方面而定。科技论文一般用短语性标题,不带感情色彩且简洁、明快,能用简单的词组就不用复杂的,尤其要慎用少用介词短语修饰。如《高温炉水中 Cl^- 对碳钢水冷壁管行为的研究》,题目语法不通,颠倒了研究者与研究对象的关系,没有表达清楚论文中"碳钢水冷壁管的腐蚀受高温炉水中 Cl^- 的影响程度"这一主题,修改后的题目为《高温炉水中 Cl^- 对碳钢水冷壁管腐蚀行为的研究》[9]。

2.3 题目的确定

确定论文题目对作者和编辑都非常重要,但因为每个研究者的方向、研究手段存在差异,不可能有现成的规律可循、现成的方法可鉴,要靠作者和期刊编辑在工作中去探索、研究、总结。《广州大学学报》编辑部的刘少华老师结合长期的编辑实践,总结出以下三条可供研究者借鉴的方法[9]。

2.3.1 按照科技论文的风格选择合适的题目类型和制式

求真务实、严谨简明是科技论文应有的风格。论文题目的类型和制式要与所投期刊的风格吻合。科技期刊编辑不应认同幽默、豪放、夸张的论文题目,因为题目的风格往往代表着论文的风格。学术论文题目大致可分为范围式、问题式、结论式、叙述式和对比式五种[10]。人文社科类论文的题目较为灵活,以上五种都有应用。对科技论文而言,问题式题目、结论式题目、对比式题目一般适用于学术讨论的论文,范围式题目多适用于综述文章,如《桩基检测技术的若干进展》。鉴于研究简报类科技论文研究问题的复杂性,一般使用叙述式题目标出对某一主题的研究,如《土塞效应对管桩低应变测试视波速的影响研究》[4]、*New interaction model for vertical dynamic response of pipe piles considering soil plug effect*[11]。

从语法角度可以将学术论文题目分为词组、单句、复句和复合四种制式。科技论文简练明达的特点决定了它一般采用词组制式。

2.3.2 按照论文正文内容的内在逻辑关系遴选若干备选题目

一篇优质论文的各部分内容均应有内在的逻辑关系。先写什么,后写什么,按照逻辑推演最后结果会是什么,中心是什么等一系列问题,只要作者把这些问题想透了,编辑理清了,就可以拟定若干个题目名称作为备选对象。以笔者发表的一篇论文为例[4],论文的主要内容是利用附加质量模型考虑土塞效应,建立了半正弦脉冲激振力作用下管桩桩顶速度响应半解析解,基于所得解详细分析了土塞效应对管桩视波速的影响。论文题目可以初步定为:《考虑土塞效应的管桩纵向振动理论研究》或《土塞效应对管桩低应变测试波速的影响研究》。进一步分析发现,该文最重要的发现是随着土塞性质的变化,管桩低应变测试波速不再是一个常数而是随着土塞高度的增大而逐渐减小,因此,为了突出论文的创新性,最后将题目定为第二个。

2.3.3 按照论文的创新成果确定最终题目

科技论文大多是研究者最新研究成果的简报,不管是理论类、数值类还是实验类,都应有所创新。科技论文的创新表现在以下几个方面。

(1)新发现。如发现新的自然现象、新的物质、新的物体和新的规律等。

(2)新理论。如新观点、新概念和新定律(理)。

(3)新的合成物质。如新材料、新配方、新产品和新结构等。

(4)新技术、新方法。

一篇论文写好以后,研究者应该比较前人的研究成果,思考几个问题:所写论文的创新点、突破点在哪里?论文有没有很好地将创新点和突破点阐述清楚?论文题目有没有涵盖所有新工作?研究者应该做到心中有数,编辑研读论文,也能作一个比较准确的判断。根据几个备选题目,结合创新点的要求,确定论文的最终题目。科技论文题目从字面上不容易吸引眼球,但如果其创新性跃上了题目,就能惹人注目、引人入胜。

2.4 题目的结构类型

英文科技论文题目的结构简单,式样繁多。它常常借助于词法、句法和修辞手段,以简明扼要、立意新颖、客观精辟的文字形式浓缩文章的基本内容,使读者能一览而知其大概,从其内容领悟出主旨。基于2004年Haggan和2007年Soler的分类标准[12,13],英文科技论文题目的表层结构可分为以下三类。

(1)名词短语结构类型。此类又可细分为以下四类[14,15]。

A. 名词短语+名词短语结构。由两个及两个以上的并列名词短语组成。

例 1　*Helical pile application in foundation improvement*

例 2　*Model test and numerical simulation on rigid load shedding culvert backfilled with sand*[16]

B. 名词短语+介词短语结构。由一个名词短语和一个介词短语组成。

例 3　*The global atmospheric circulation on moist isentropes*[14]

例 4　*Structural patterns in empirical research articles*[15]

C. 分词短语结构。由动词的现在分词或动词过去分词+名词短语组成。

例 5　*Grouting technology in pile foundation engineering*

例 6　*Testing the speed of spooky action at a distance*[14]

D. 介词短语结构。即题目的第一个单词为介词。

例 7　*On the spontaneous emergence of cell polarity*[14]

例 8　*Towards a linguistics of new production*[15]

(2)完整句子结构类型。指的是从英语语法角度来看,其结构是符合句子构成的题目。

例 9　*Is air pollution causing landslides in China?*

例 10　*PC program estimates BTEX*[18]

上述两个例子都是完整的句子,可能是因为句子较短的缘故,显得不是很累赘。还有一点值得注意的是,两个例子采用的时态都是一般现在时,这是科技论文的一大特征。一般现在时既可以表示现实存在的事实和现在发生的事情,又可以给读者一种直接感、同时感。

(3)复合结构类型。指由冒号、句号、问号或破折号等标点符号将主、副标题连在一起的题目。

例 11　*Cell type—specific loss of BNDF signaling mimic optogenetic control of cocaine reward*[14]

例 12　*Geochemistry of geothermal waters from the Gonghe region, Northwestern China: implications for identification of the heat source*[19]

需要说明的是,复合结构类型的题目也可以称之为正副标题型题目[20],其构成一般是由前后两部分组成的,而且正标题和副标题之间多用冒号或破折号或其他标点符号隔开,两者是同位关系。一般而言,副标题不仅可以对正标题进行补充解释,而且还可以对正标题进行进一步的描述,如交代论文研究背景,或对正标题进行递进说明。

2.5 实例介绍

一个好的标题可以让同行们清楚地明白你的研究内容,即你做了哪些研究工作。标题通常用说明格式书写,而不用完整的词句。题目应简短、精练,言简意赅地表达文章的中心思想。要删除一切不必要的字词,以突出主题。

常见有缺陷的标题实例及修改如下。

例 1 *Synthesis of a contrast agent for MICP of calcareous sand in the South China Sea*

点评:上述标题没有说明合成了什么样的对比剂。

建议:*Synthesis of a contrast agent with Pasteurella for MICP of calcareous sand in the South China Sea*,这样就体现了合成对比剂的主要成分为巴氏杆菌,更容易吸引相关读者查阅。

例 2 *Effect of soil plug on AWVPP*

点评:读者无法明白 AWVPP 是什么,也不明白土塞效应对 AWVPP 具体有什么影响。

建议:*Effect of soil plug on the changing mechanism of apparent wave velocity of pipe pile*(*AWVPP*),这样明确了 AWVPP 即为管桩视波速,也体现出土塞的存在会使管桩视波速产生变化。

例 3 *Study of concrete in dam*

点评:此标题没有说明用速凝混凝土对大坝做了什么研究,研究的目的是什么,及研究用的是什么样的混凝土。

建议:*Study of the stability of dam by using rapid concrete*,这样明确了研究目的是使用速凝混凝土来研究大坝的稳定性。

例 4 *A study of magnetic resonance imaging in rock*

点评:此标题没有表达出用磁共振在岩石方面做了什么研究。

建议:*A study of magnetic resonance imaging in the detection of rock joint plane*,这样就明确了是利用磁共振来检测岩石节理面的。

例 5 *Analysis of the vertical free vibration characteristics of pipe pile with exposed cushion cap considering soil plug effect*

点评:此标题字数较多,显得冗余。

建议:*Vertical free vibration characteristics of pipe pile with exposed cushion cap*,该文的核心思想是考虑了承台效应,只要保留关键信息即可。

本章参考文献

[1] 拉姆奇.如何查找文献[M].廖晓玲,译.北京:北京大学出版社,2007.
[2]《兰化科技》编辑部.科技论文标题[J].兰化科技,1989,7(3/4):234-236.

[3]张润中.科技期刊论文的标题与主题[J].编辑工程,1989,1(3):152-156.

[4]刘浩,吴文兵,蒋国盛,等.土塞效应对管桩低应变测试视波速的影响研究[J].岩土工程学报,2019,41(2):383-389.

[5]黄永松.科技论文标题的确定[J].安徽教育学院学报,2000,18(6):92-93.

[6]李胜春.科技论文标题的实例分析[J].重庆大学学报(社会科学版),2001,7(4):66-68.

[7]杨惠美,刘国生.怎样拟定科技论文标题[J].东北林业大学学报,1987,15(6):85-88.

[8]张志,郑洪生.科技论文标题的撰写[J].防护林科技,1997(4):46-47.

[9]刘少华.科技论文标题的要求与确定方法[J].中国科技期刊研究,2008,19(3):498-500.

[10]马东震.简论学术期刊的论文标题[J].宁夏大学学报(社会科学版),1997,19(1):123-126.

[11]WU W B,El NAGGAR M H,ABDLRAHEM M,et al. New interaction model for vertical dynamic response of pipe piles considering soil plug effect[J]. Canadian Geotechnical Journal,2017,54(7):987-1001.

[12]HAGGAN M. Research paper titles in literature, linguistics and science: dimensions of attention[J]. Journal of Pragmatics,2004,36:72-75.

[13]SOLER V. Writing titles in science: an exploratory study[J]. English for Specific Purposes,2007,26:124-137.

[14]曹杨,赵硕.科技论文标题的结构和语言特征:以 *Science* 和 *Nature* 为例[J].外语教学,2014,35(2):35-39.

[15]赵聪超.英语语言学权威学术论文题目结构和语言特征的对比研究[J].兰州工业学院学报,2017,24(2):137-141.

[16]CHEN B G,SONG D B,MAO X Y,et al. Model test and numerical simulation on rigid load shedding culvert backfilled with sand[J]. Computers and Geotechnics,2016,79:31-40.

[17]ZHANG M,MCSAVENEY M. Is air pollution causing landslides in China?[J]. Earth and Planetary Science Letters,2018,481:284-289.

[18]刘洪泉.科技论文的标题结构浅析[J].上海科技翻译,1997,1:16-19.

[19]LIU M L,GUO Q H,ZHANG X,et al. Geochemistry of geothermal waters from the Gonghe Region, Northwestern China: implications for identification of the heat source[J]. Environmental Earth Sciences,2016,75:682.

[20]陶坚,黄大网.基于语料库的学术期刊论文英文标题对比研究:学科及文化的双重维度[J].当代外语研究,2010,9:33-35.

第 3 章　摘要与关键词

3.1　概　述

摘要是论文的浓缩。它以简单易懂的文字直接陈述论文的内容,不加注释和评论,清楚而完整地显示出论文的概貌,通常出现在标题和作者之后、关键词之前。摘要应具有自明性(sefl-contained)和独立性(standing on one's own)[1],并拥有与一次文献同量的主要信息,即不阅读全文就能获得必要的信息。很多读者往往会通过阅读论文摘要来判断是否是他要找的论文。因此,摘要的准确性、规范性直接影响到论文在国内外数据库中的收录率和索引率,间接影响着论文的传播效果[2]。

关键词是从论文标题和正文中选取出来的表现论文主题内容、具有关键作用的规范化单词或术语,单独标在摘要之后、正文之前,其作用是表示某一信息数目,便于文献资料和情报信息检索系统存入存储器,以供检索。读者通常通过检索关键词来进行文献检索。因此,一篇论文关键词的合理选取将直接决定该论文能否进入读者的视野[3]。

3.2　摘要的定义和功能

摘要是论文的缩小版,应该提供正文的简单小结并全面涵盖论文的每一个部分:前言(introduction)、材料与方法(materials and methods)、结果(results)、讨论(discussion)。工科类专业的科技论文主要有实验类、数值模拟类、理论分析类、工程案例分析类和综述类五类。总的来说,不管是哪一类论文,其摘要主要有三种类型:报道型(informative abstract)、指示型(indicative abstract)及这两种类型的结合型,即报道-指示型(combination of informative and indicative abstract)。报道型摘要完整地报道原文中的具体内容,特别是研究或实验报告中的研究结果、结论等。指示型摘要亦称概述性摘要,一般只对全文作扼要叙述。报道-指示型摘要是把论文的主要方面写成报道型而将其次要方面写成指示型。

好的摘要使读者能够迅速而准确地获取论文的基本内容,判断该论文是否与自己的研究兴趣相关,从而决定是否需要阅读全文。摘要的内容应包括四个要素,即目的、方法、结果、结论。

(1)目的。指研究的前提和缘起,即为什么要做此项研究,可以有简单的背景材料。

(2)方法。指研究所用的原理、对象、观察和实验的具体方法等。

(3)结果。指研究的结果、效果、数据等,着重反映创新性的、切实可行的成果,包括本组研究中的重要数据。

(4)结论。指对结果进行综合分析、逻辑推理后得出的判断。有的可指出实用价值和推广价值;如有特殊例外的发现或难以解决的问题,可以提出留待今后深入探讨。

此外,一篇好的英文摘要还需具有完整、精练、准确、正规的特点[4]。完整指的是摘要本身以文字的形式将论文的主要内容全面地反映出来,不能用插图、表格或者文中的某个标题代替说明。精练是指摘要必须简明扼要地概括论文的精髓,不宜包含与其他研究工作对比、列举例证等内容,篇幅长短以250个实词以内为宜(《美国工程索引》要求它录入的摘要每篇不能超过150个实词[5])。准确指的是既要忠实于原文,又要遵循语言表达习惯。正规说的是论文摘要属于正式文体,用词、语法、句法、结构等行文要求都必须严谨、规范。

科技论文英文摘要具有六大功能:一是报道功能,它能准确传达最新的学术成就;二是交流功能,它是国际科技交流的重要工具;三是检索功能,读者通过它可以找到需要的文献;四是参考功能,读者可以从摘要中得到有用的信息;五是信号功能,指的是摘要发表的及时性;六是示址功能,提供文献的原始出处、作者等[6]。

3.3 摘要的规范化写作

英文摘要的文体特征主要体现在语篇、句法、词汇和语相这四个层面。语篇的特征主要指结构上的衔接性、语义的连贯性、意义上的完整性、句子间的逻辑性等;句法层面主要涉及时态、语态和人称,句型结构等;词汇层面要求使用规范、准确的词汇;语相层面包括字体、大小写、标点、文字符号的排列形式等。由于语相层面的内容相对简单,这里主要从语篇、句法、词汇三个方面进行阐述[4]。

(1)英文摘要作为独立的语篇,是一个上下连贯、前后一致、条理清晰的语言整体,其实际目的通过发挥各个语步的不同功能得以实现。语篇应围绕论题形成一个完整的结构,语篇中的句子之间存在一定的逻辑关系并在此结构基础上通过衔接成分组合在一起,从而使这个连贯的语篇符合认知、语用和语义的原则。英文摘要的语篇虽然短小,但仍然应该按照主题句(研究目的、方向)、拓展句(研究方法、结果等)和总结句(研究结论、启发等)这种英语语篇的方式展开,层层推进,完成写作[8]。

(2)英文摘要的句法层面涉及时态特征、语态特征和人称的选择等方面[9]。

A. 时态特征。以简练、准确为佳,主要使用一般过去时和一般现在时,较少使用现在完成时、过去完成时和一般将来时,其他时态基本不用。

B. 语态特征。摘要可以使用被动语态,也可以使用主动语态。在20世纪六七十年代以前,被动语态成为科技论文英文摘要的一大特色,占绝对优势,但随着时间的推移,主动语态逐渐占据主流。SCI(Science Citation Index,科学引文索引)和EI(Engineering Index,工程索引)等国际著名科技文献检索机构在编写格式方面也规定:尽量用主动语态替代被动语态。

C. 人称的选择。以前,我国的编辑和作者在加工和撰写英文摘要时都习惯于使用第三人称。20世纪70年代后,很多国际期刊开始倡导使用第一人称进行科技论文的写作。EI

的写作要求中也多次提出使用第一人称"we"(我们)可以拉近作者和读者之间的距离,使读者产生共鸣,也使文章更具可读性[7]。目前,大多数国际期刊都提倡使用第一人称"we"(不能使用"I"),比如《自然》(Nature)、《细胞》(Cell)等。

(3)英文摘要的词汇使用原则是规范、简洁[9]。在一般情况下,应该用合适的单词替代短语,用合适的短语替代句子,比如"提取"一词用"extract"替代"take out";有的词语义项较多,比如"come out"一词,有"出版""开花""上市"等意,为避免产生歧义,宜回避使用;在复合句中,通常用动词不定式短语、分词短语、独立主格结构、介词短语等形式替代状语从句,以达到言简意赅的效果。

接下来,从语篇图式结构和语言规范两个方面对英文摘要规范化写作进行论述。

3.3.1 符合英文摘要的语篇图式结构

了解和掌握英文摘要的体裁特征是写出规范化语篇的重要前提。根据斯威尔斯(Swales)的体裁分析理论,以语步为语篇分析的出发点,结合摘要语篇三种类型的特点,具备常规语篇要素的英文摘要就体现出引言—方法—结果—结论/讨论四语步模式和引言—方法—结果或方法—结果—结论三语步模式[10]。英文摘要的体裁结构由引言、方法向结果、结论步步推进,形成了语篇的宏观图式结构(schematic structure)。由于英文摘要要求结构清晰、用词准确,因此掌握一些常用词汇、句型和规范的表达方式是非常有必要的[11]。

1. 引言语步

引言语步主要介绍研究的前提、目的、任务或引入论题等,比如表述研究目的时,可用:

(1)名词。常用的名词有 goal、aim、purpose、objective 等。带有名词的典型句型也有很多,比如,本文旨在……(The paper is to …),此项调查的主要目的是……(The primary objective of the survey is to …),等等。

(2)动词。常用的动词有 aim at、intend to、attempt to 等,例如,本文旨在查明……(The article is aimed at finding out …)。此外,动词不定式短语因其简洁明了的特点,也常常用在目的语步中,例如,为确定……(To determine …),为鉴别……(To identify …),等等,但需要注意的是,当使用动词不定式表示研究目的时,主句部分必须有主语。

2. 方法语步

方法语步说明了为实现研究目的而运用的理论、材料、手段、程序等或简介研究过程。

(1)应用了某种理论、方法、某种材料等。其常用的名词有 application、adoption、use 等,比如,应用这种新理论是为了……(The application of the new theory is to …);常用的动词有 apply、use、explore、adopt 等,例如,用卡方检验法评估这次观测……(The observations were assessed using the chi-squared test …)。

(2)在进行实验描述时,test、observation、sample、experiment 等名词经常出现;observe、test、monitor、conduct、carry out 等动词常被用到,例如,使用……做试验(The test was conducted by using …)。

3. 结果语步

结果语步阐释运用上述方法得到的成果。描述研究结果常用的名词有 findings、results、outcome 等，常用的动词有 indicate、suggest、show、provide、demonstrate、present 等，例如，获取数据表明……（The data obtained suggested that…）。

4. 结论语步

结论语步的内容是通过对研究成果进行分析、评价而推出的判断或提出的问题、建议。提出结论常用的名词有 conclusion、summary 等，动词有 conclude、draw、reach、arrive at、come to 等，例如，可以得出以下结论……（The following conclusion can be drawn/reached…），通过试验我们得出结论……（We concluded by the test that…）。

3.3.2 符合英语语言写作规范

在语言运用方面，英文摘要应该符合英语的写作规范。这里主要从语篇、句法、词汇等层面论述。

1. 语篇层面

要从语篇层面使英文摘要符合英语语言写作规范，衔接（cohesion）和连贯（coherence）是需要把握的核心[12]。

衔接是指整个语篇中语言成分之间的语义关联。衔接的实现手段主要有五种：①照应（reference），包括运用指示代词（比如 this、that、these、those、here、the 等）形成的指示照应和运用人称代词（比如 he、she、they、their、we 等）形成的人称照应及运用具有比较意义的词（比如 compare、different、more、other 等）形成的比较照应；②省略（ellipsis），指名词等词类的省略；③替代（substitution）；④连接（conjunction），指用副词、连词和词组把语义连接起来；⑤词汇衔接。

连贯的实现手段通常是运用语境推理（using context of situation）、想象（using imagination）等。

衔接是篇章的有形网络，涉及语篇的结构形式；连贯是篇章的无形网络，源于篇章的语义内容。在英文摘要的成文过程中，把握衔接和连贯两个核心，运用其实现手段，依照英语的表达规范，实现信息交际功能。

2. 句法层面

由于东西方的文化差异和思维方式差异，摘要的语言表达方式在句法层面的要求是结构规范。

1）语序

语序是指句子内部的结构顺序。英文摘要以介绍和叙述为主，多使用自然语序的、主谓结构的陈述句[13]。一是主语和谓语负载着句子中最重要的信息，是句子的核心，能够突出摘要的内容；二是这种结构比较紧凑，符合科技论文的严密性。另外，在英文摘要中，当某些成分比较复杂、冗长时，就涉及调整语序或换序的问题。比如某些摘要中被动语态句子的主语很长，呈现出头重脚轻之势，给读者一种透不过气的感觉，这一点极不符合英文的行文习惯。如果调整语序,添加主语"we",改为主动语态,则效果要好得多。为使句子结构紧凑规范,对于有多个限定词修饰的短语,应避免一长串名词或形容词罗列在中心词前面,把这些修饰成分分解为前置或后置成分,或者用连字符将其紧密连接。

2)时态

英文摘要的时态选择主要受交际目的和各语步交际功能的影响,在各语步中的分布是有规律的[14]。

(1)引言语步通常使用一般现在时,以表述研究目的、任务或此项研究的重要性;如果背景介绍涉及他人已经取得的成果或得出的结论,这时应该结合使用现在完成时。

(2)方法语步多用一般过去时,描述作者已经使用过的理论、手段、材料或者介绍其研究过程。

(3)结果语步经常使用一般过去时,陈述试验或研究的成果。在某些理论性或语言学、数学类论文中,结果语步倾向使用一般现在时,因为作者认为其研究结果与时间关联不大。

(4)结论语步常选用一般现在时来提出建议或提出问题。

总体而言,时态的选择还应视具体情况,采取灵活变通的方法。

3)语态

语态的正确使用是英文摘要写作的基本要素之一。被动语态的使用在早期的英文摘要行文中是非常普遍的,这样可以避免提及动作的执行者,去除主观成分。近年来,我国的广大科技论文作者由于缺乏与国际科技编辑界的最新信息交流,对英文摘要中的语态使用方面的认知仍停留在传统观念上,认为英文摘要显著的句法特征就是大量使用被动语态,结果导致被动语态滥用的情况比比皆是,这种滥用会使句子结构笨拙呆板,语义模糊不清。针对以下情况,使用主动语态会有更好的文体效果。

(1)语篇标志方面。在引言语步的最后一句或方法语步的第一句,使用语篇标志词"we"引导主动语态的句子,可以起到很好的承上启下作用和强调作用;在结果语步使用"We find that ……"或者在结论语步使用"We conclude that ……"或"We recommend that ……"等这类明确表明各语步主动语态的句子可以使语言层次分明,语义更加连贯。

(2)突出信息表达方面。在英文摘要中,位于主位(theme)的主语通常是已知的信息,位于主位之后的述位(rheme)是未知的信息,由于信息的核心内容集中在述位,所以述位的语篇价值要高于主位。这就是句尾焦点原则(end focus)。在这种情形下,国内作者偏爱使用的被动语态基本起不到突出信息的作用。例如,Twenty isolates were detected with no mutation of the gene,此课题研究的重点是观察这种基因是否产生突变,如果用"gene"作主语,"mutate"作谓语,句子使用主动语态,则会有信息突出、一目了然的效果,改为:The gene did not mutate in twenty isolates。当然,主动语态与被动语态没有绝对的孰优孰劣,只要能够清楚准确地表达摘要的信息即可。

3. 词汇层面

在英语词汇层面,英文摘要的规范化写作主要是指书面语使用和专业词使用这两个方面[15]。

(1)使用书面语。与英语中日常使用的表示事物、行为、现象、性质等通俗易懂、自然简练的通用词语不同,书面语同理论与逻辑密切相关,基本没有感情色彩,所以在撰写英文摘要时,需从表现力和意义方面分析语言手段,恰当得体地使用书面语,比如,get→obtain①。至于缩略语,除了本专业或临近专业科研人员公知公用的词汇(如 DNA、GPS、CAD 等),在一般情况下,英文摘要不使用新闻报道或生产生活中常用的缩略语,例如,ad→advertisement、auto→automobile、mike→microphone 等。

(2)使用专业词汇或学术词汇。就词源而论,大多数专业词汇或学术词汇来源于拉丁文和希腊文,虽经历了时代的不断变迁,但两种语言的词义、词形基本没有什么变化,也不易产生歧义。选择这样的词汇会使语篇概念准确、推理严密,提升语篇的严谨性。

3.4 关键词的定义与功能

关键词(keywords 或 key words)是文献题名中的主要词语,或是作者及编辑认定为该文献的中心词语[16,17]。《现代汉语词典》的定义为:①能体现一篇文章或一部著作的中心概念的词语;②指检索资料时所查内容中必须有的词语[18]。规范的关键词有利于该篇论文的被检索和被引用。国家标准《科技报告编写规则》(GB/T 7713.3—2014)中明确规定:"每篇报告、论文应选取 3~8 个词作为关键词,……为了国际交流,应标注与中文对应的英文关键词。"[19]

关键词在科技论文的文献检索中是最重要的标识。一般规范的关键词是从科技论文的题目、正文中的主要内容、重要的层次标题中精心挑选出来的,必须能反映文章的主题。学术界常利用关键词去检索最近发表的论文,文献数据库也常根据关键词检索和收录相关的文章。如果关键词不规范不恰当,则论文被检索和收录的概率一定会很低。有些作者在英文关键词的选用上有一些不当之处,比如,有些作者选择了与文献中心不相关的词或词组,或选择了文献中没有参考价值的概念词。这样会造成文献检索的错检和漏检。又比如,有的在选择关键词时未选用专业规范的词,而是将非公知的短语、词汇,甚至有些缩略语作为关键词。只要有公知的专业术语,就应该使用专业术语,不可以自行想当然地选择。

3.5 关键词的合理选择与使用

3.5.1 关键词的优化

由于英文关键词选取在一定程度上受到全禁用词(full-stop words)、半禁用词(semi-stop words)等规定限制,为此,在其词语的取舍上不妨采取以下优化措施。

1. 避用全禁用词

被 SCI 称为检索词(关键词)的全禁用词的词类有冠词、介词、连词、代词、副词、形容

---

① 表示不用 get,用 obtain;后同。

词、感叹词、某些动词(连系动词、情态动词、助动词)等[20]。从检索的角度来讲,上述词类用作关键词无实质意义。从词性上分析,不难看出,英文关键词通常用名词或名词词组形式。

在标引英文关键词的时候,尽量不要使用 and、of、&、—、* 等连接词、介词和符号,也要避免使用冠词[21]。谈及连词,例如,"energy saving and emission reduction"(节能减排)这一关键词最好拆分为"energy saving"(节能)和"emission reduction"(减排)两个关键词,这样既避免使用了连词"and",同时又体现了关键词的专指性。提到介词,例如"influence of native language"(母语影响)这一关键词就可改为名词修饰名词的短语形式"native language influence",而"figure of speech"(修辞)这一关键词可选择其相同意义的名词"rhetoric"(修辞)来表达。至于冠词,尽管英语语法规定在某些情况下,如序数词、最高级形容词前须加定冠词,但在作为关键词列出时,定冠词一概略去[22]。例如,the United States(美国)、the late Qing Dynasty(晚清)、the fifth media(第五媒体)等用作英文关键词时,其定冠词 the 都应省去。

2. 活用半禁用词

半禁用词是指那些不能反映论文主题,但可以配合检索作用的词语。国际著名的数据库系统,如 SCI、EI、SDOL(Science Direct On Line)等,把诸如理论、报告、实验、研究、方法、分析、问题、途径、特点、目的、概念、规律、发展、现象等词语,视为无特殊检索意义的半禁用词[23]。此类半禁用词过于空泛,不宜独立用作关键词。"由于这些关键词在大多数学术论文中几乎都可以使用,它们在提示具体某一篇论文主题内容的专指性方面就大大降低,失去该关键词应起的基本作用。"[24]它们可以加以变通,与其他半禁用词或实义词组成关键词。

以下英文关键词都是对半禁用词加以优化组合而成的:relevance theory(关联理论)、reform experiment(改革实验)、comparative study(比较研究)、compiling methods(编纂方法)、pragmatic analysis(语用分析)、psychological problems(心理问题)、implemental way(实施途径)、interpretation characteristics(口译特点)、developmental pattern(发展规律)、diachronic development(历时发展)、phenomenon research(现象研究)等。

3. 弃用句子

句子(sentence)是具有主语部分和谓语部分并有完整意义的、可以独立出现的一组词。就英文关键词而言,不可用英文句子。有时为了表达与中文关键词相同的意思而不得不用含有介词的名词短语,但最好要做到名词短语中尽可能少地含介词或连词。非用短语不可的,短语也不宜太长,只要标引的关键词能说明问题,用词量越少越好[21]。

3.5.2 关键词的形式

英文关键词的形式,如大小写、单复数、缩略语等的表述应遵循英语语法的相关规定,并符合文献检索的要求。

1. 大小写

英语语法规定专有名词要大写。专有名词是特定的人、地方及机构或团体的名称。英文关键词中主要涉及专有名词的有人名、地名、事物名称,以及公知公认的缩略语或专业术语等。除此之外,英文关键词基本上都用小写字母排版,当首字母大小写均可时,一律小写[22]。需说明的是,英文关键词的大小写有时因各个刊物的要求不同而相异。

2. 单复数

既然英文关键词采用名词或名词词组的形式,就必然会涉及其单复数这一问题。关键词基本上都采用单数形式,但有些事物通常以复数状态存在或出现,这时就必须用复数[22]。就名词而言,可分为可数名词和不可数名词,仅可数名词有单复数之分。在可数名词用作英文关键词的问题上,美国地理学家协会(Association of American Geographers)主张采用其复数形式并举例:若在 race 或 races 中加以选择,宜选用后者[25]。

3. 缩略语

英语缩略语覆盖面广,涉及政治、法律、历史、经济、语言、文化、军事、教育等各个领域。由于它的语用优势及丰富的表现形式,得到了广泛应用。然而,在科技论文英文关键词撰写中,切不可随心所欲地任意使用英文缩略语。作为关键词,外文缩略语应用全称形式以免产生歧义[25,26],例如,open access(开放存取)、sodium chloride(氯化钠)、problem-based learning(基于问题的学习)和 graphic process unit(图形处理单元)等都不宜用其缩略语 OA、SC、PBL 和 GPU 作为关键词。AIDS(艾滋病)可以直接用作关键词。

尽管有多种解释意义的外文缩略语不能被直接用作关键词,但某些全拼形式特别长的或者缩略语形式比其全称更为人所知的,如人们对 DNA(脱氧核糖核酸)远比对其全称 deoxyribo nucleic acid 熟悉得多,可以直接用作关键词。有机化合物 NADP(烟酰胺腺嘌呤二核苷酸磷酸)是 nicotinamide adenine dinucleotide phosphate 的缩写,该缩写已作为词条列入《韦氏词典》,也可以直接使用。

3.5.3 关键词的排序

英文关键词的正确排序不仅能清晰地反映该论文的主题内容,而且能提高其论文的被检索率、被阅读率和被引用率。正确的排序应该基于主题内容的等级层次(hierachy)[17],依据词语之间的逻辑联系。

1. 等级层次

关键词排序的等级层次是指关键词按论文主题内容由高到低、由大到小的层次进行排序。从等级层次上讲,首标关键词的等级层次应该是最高或最大的,最能反映该学术论文的最主要内容。

不论是由高到低还是由大到小,等级层次都应反映出主题范围、研究方法、具体内容和有利于检索和文献利用的其他关键词。此排序法既遵循 Westburn Publishers Ltd. 对英文关键词的排序要求[27],从中、英文关键词一一对应的角度讲,也符合中国科协学会学术部在的规定[28]。

2. 逻辑联系

逻辑联系是指关键词按词语之间由浅入深、由点到面的联系进行排序。各关键词之间要有一定逻辑关系,即选词和排列次序要反映出词与词之间的逻辑联系[29]。

3.6　实例分析

现以笔者三篇英文论文的摘要和关键词为具体实例,说明英文摘要和关键词的写作方法。

实例一　理论分析类论文[30]

Abstract:The apparent phase velocity of open-ended pipe piles after installation is difficult to predict owing to the soil-plug effect. This paper derives an analytical solution to calculate the apparent phase velocity of a pipe pile segment with soil-plug filling inside (APVPSP) based on the additional mass model. The rationality and accuracy of the developed solution are confirmed through comparison with the solution derived using the soil-plug Winkler model and experimental results. A parameter combination of the additional mass model that can be applied to concrete pipe piles used most commonly is recommended. The attenuation mechanism of the soil plug on the APVPSP is clarified. The findings from this study demonstrate that the APVPSP decreases with the mass per unit length of the pile,but has nothing to do with the material longitudinal wave velocity of the pipe pile. The APVPSP decreases significantly as the impulse width increases;however,for pipe piles without soil-plug filling inside, the impulse width has negligible influence on the apparent phase velocity.

Key words:Apparent phase velocity, soil plug, pipe pile, additional mass model,low strain test.

第一句介绍研究的背景。
第二句和第三句介绍采用解析解开展研究,并通过与已有解和试验数据进行对比验证。
第四句和第五句介绍研究结果。
第六句介绍研究结论。

关键词分析:这五个英文关键词按由高到低的等级层次排序,第一个关键词"Apparent phase velocity"的等级最高,为研究目的;"soil plug"和"pipe pile"的等级次之,为研究对象和因素;"additional mass model"的等级再次之,为研究方法;"low strain test"的等级最低,为研究应用领域。这种由高到低的等级层次排序能层层深入地反映论文主题。

实例二　数值模拟类论文[31]

Abstract:A DEM simulation of the installation process of open-ended pile is conducted by means of the particle flow code PFC 2D. Focus is placed on the investigation of the

第一句介绍研究的任务和引入主题。
第二句~第四句介绍数值模型建模方法及验证方法。

soil plug behavior, from both macroscopic and microscopic perspectives. A soil assembly with natural initial stress state is first generated and the validity of the numerical installation process is further checked. Afterwards, a macroscopic analysis is performed based on the porosity and stress state. Numerical results indicate that a dense zone, with the length about half the pile diameter, has formed at the pile tip. A dramatic load transfer is observed at the bottom of the soil plug, where the lateral pressure coefficient peaks at around 2.4. From a microscopic perspective, two modes of soil layer deflection exist for soil mass inside and under the pile, respectively. Different particle displacement patterns indicate that with increasing penetration depth and resistance accumulation, soil mass flows into the pile with a lower rate. Further analysis based on the distribution of contact force chain and principal stress rotation show that soil particles, when subjected to external disturbance, tend to rearrange to the most stable structure, which is in the arch shape. Finally, an improved arch model based on the numerical results is proposed to facilitate the understanding of the plug behavior.

Key words: DEM, open-ended pile, sand plug, principal stress rotation, arching effect.

实例三　工程案例分析类论文[32]

Abstract: Reinforced concrete (RC) culverts under high fill have been widely used in the construction of expressways and railways. Based on the results of a field survey, various types of structural damage to RC culverts occur during construction and service periods. In this study, numerical simulation and field tests were conducted to investigate the impact of ground soil properties on the structural integrity of RC culverts under high fill. Important factors that influence culvert integrity, such as ground bearing capacity and ground treatment, have been analyzed in detail. Research findings indicate that damage to RC culverts under high fill is not typically caused by failure to the subgrade layer under the culvert foundation, due to the beneficial effects of foundation recess depth, foundation width and subgrade layer consolidation. Structural damage is probably caused by improper ground treatment strategies. Proper strategies for preventing integrity problems are recommended based on the research.

Key words: culvert, structural integrity, high fill, ground bearing capacity, ground treatment.

整性对地基承载力和地基处理的影响,范围进一步缩小。

通过上述案例分析可以看出,英文摘要是科技论文的缩影,大量的信息被充分压缩后,包容在字数极其有限的语篇内。每个单词、短语、句子的信息承受量极高。撰写和翻译英文摘要的基本原则即 ABC 原则:accuracy(准确)、brevity(简洁)、clarity(明晰)。本着科学严谨的态度,把握其体裁的语篇图式结构特点,了解其文体特征和写作规范,加强英文素养,对读者和作者掌握英文摘要谋篇布局的机制、遣词造句的技巧,实现英文摘要的规范化,促进科研成果的交流和推广,很有裨益。

同时,掌握好学术论文英文关键词的选择与使用技巧:在词的优化上,应避用全禁用词,活用半禁用词,弃用句子;在表述形式上,应正确使用大小写、单复数、缩略语;在排序方式上,要基于主题内容等级层次、依据词语之间的逻辑关系。此举无疑能促进英文关键词撰写的规范化,并有助于该学术论文的检索和引用。

本章参考文献

[1]邓凌.学术论文英文摘要翻译中的常见问题浅析[J].青海师范大学(哲学社会科学版),2008(6):120-121.

[2]冯恩玉,吴蕾.国内外光学类科技期刊论文英文摘要体裁对比分析[J].中国科技期刊研究,2016,27(2):230-236.

[3]朱虹.科技论文的英文题目、摘要及关键词的编辑与加工[J].黑龙江教育学院学报,2004,23(2):155-156.

[4]李涛.科技论文的英文摘要规范化问题研究[J].辽宁工业大学学报(社会科学版),2018,20(6):70-73.

[5]金丹,王华菊,李洁,等.从《EI》收录谈科技论文英文摘要的规范化写作[J].编辑学报,2014,26(S1):S118-S120.

[6]刘英丽.如何编写科技文章的中文摘要[J].农业图书情报学刊,2006(9):14-15.

[7]周强,侯集体,张立春.科技期刊国际化视野下英文摘要分析[J].中国科技期刊研究,2016,27(7):804-810.

[8]陈欣,闫清.科技论文英文摘要的语篇结构对比研究[J].科研园地,2016(2):13-16.

[9]张菊.科技论文的英文标题、摘要及关键词的翻译技巧[J].科技信息,2008(12):24,61.

[10]林江娇.浅议科技论文结构型英语摘要的句型特点和表达方式[J].江苏理工学院学报,2017,23(3):107-110.

[11]陈竹,李洁,王华菊,等.科技论文英文摘要的写作[J].编辑学报,2016,28(S2):13-17.

[12]李芳萍,李舰君.科研论文英文摘要中的衔接:以船舶与海洋工程为例[J].哈尔滨职业技术学院学报,2015(6):110-111,153.

[13]杨玉华.科技论文标题、关键词及摘要的撰写与英文翻译[J].焦作大学学报,2009,4(2):34-36.

[14]孙文豪.土木工程论文英文摘要的写作实践与原则分析[J].海外英语,2015(7):134-136.

[15]赵常友,奚丽云.英文学术期刊论文标题、摘要及关键词词汇模式研究[J].重庆教育学院学报,2012,25(4):89-92.

[16]QUITMAN I L. Simon & Schuster quick access:reference for writers[M]. Toronto:Pearson Prentice Hall,2007:158.

[17]王伟.学术论文英文关键词撰写的技术性细节[J].湖北第二师范学院学报,2013,30(9):122-126.

[18]中国社会科学院语言研究所词典编辑室.现代汉语词典[M].6版.北京:商务印书馆,2012:477.

[19]全国文献工作标准化委员会.科技报告编写规则:GB 7713—2014[S].北京:中国标准出版社,2014.

[20]刘大乾.SCI关于关键词的一般选取准则及词义库建设[J].中国科技期刊研究,2007,18(6):1073-1074.

[21]战英民.关于文献关键词标引问题[J].河北科技图苑,2010,23(4):19-25.

[22]史成娣,钟传新,杭桂生.科技论文中英文关键词的规范表达[J].中国科技期刊研究,2005,16(6):919-920.

[23]杜香莉,王立宏,罗红彬.我国期刊全文数据库关键词规范化问题探讨[J].中国科技期刊研究,2007,18(4):612-614.

[24]顾泉佩.学术论文中关键词的合理使用[J].中国科技期刊研究,2003,14(1):102-103.

[25]Association of American Geographers. Abstract guidelines[EB/OL]. [2020-05-10]. https://aag-annualmeeting.secure-platform.com/a/page/abstracts/abstract-guidelines.

[26]陈航,黄春杨.学术期刊论文关键词的规范化问题[J].航海教育研究,2006,23(1):110-112.

[27]Westburn Publishers Ltd. Abstracts & keywords[EB/OL]. [2020-04-29]. https://www.westburn-publishers.com/journals/social-business/authors/academic-papers/.

[28]佚名.中国科协《关于在学术论文中规范关键词选择的规定(试行)》[J].系统工程,2003,21(6):96.

[29] 胡蓉. 学术论文关键词探析[J]. 四川职业技术学院学报,2004,14(3):120-122.

[30] WU W B,LIU H,YANG X Y,et al. New method to calculate the apparent phase velocity of open-ended pipe pile[J]. Canadian Geotechnical Journal,2020,57(1):127-138.

[31] LI L C,WU W B,EL NAGGAR M H,et al. DEM analysis of the sand plug behavior during the installation process of open-endedpile[J]. Computers and Geotechnics,2019,109:23-33.

[32] CHEN B G,SUN L. The impact of ground soil properties on the structural integrity of high-fill reinforced concrete culverts[J]. Computers and Geotechnics,2013,52:46-53.

第 4 章　引言与研究背景

4.1　概　　述

"A bad beginning makes a bad ending."这是一句广为流传的英文谚语,好的开端是成功的基础。英文科技论文撰写也存在着这样一个总起全文决定文章质量的开端——引言。引言(也称前言、序言或概述)经常作为科技论文的开端,提出文中要研究的问题,引导读者阅读和理解全文[1]。引言作为论文的开场白,应介绍论文的写作背景和目的,以及相关领域内前人所做的工作和研究的概况,说明本研究与前人工作的关系、目前研究的热点、存在的问题及研究意义,引出本文的主题给读者以引导。引言也可以点明本文的理论依据、实验基础和研究方法,简单阐述其研究内容,三言两语预示本研究的结果意义和前景,但不必展开讨论。

总之,引言可以帮助读者在无需翻阅其他研究文献的情况下,了解该项研究的同行进展及该文研究成果的意义。不仅如此,引言能起到总领全文的作用,所提及的问题是要解决的核心,其列举的前人研究是文章主体部分的理论前提。同时,引言也会对本文主体部分所撰写的研究进行必要的介绍,在联系前人研究的同时,突出说明文章的创新性和应用性。

4.2　引言撰写的基本要求

在引言部分,作者需要说明撰写文章中所涉研究的研究背景、研究意义,以让读者对该项研究产生认可,同时需要列举前人在类似研究中的进展,方便读者建立相似研究间的递进关系,以使文章中的成果更为读者所接受。引言撰写的基本要求有以下几点[2]。

(1)在引言中,对研究问题实质和范围的描述需要放在首要地位,用最准确、精练的语言来描述文章的研究背景和研究意义。

(2)引用与该文最相关、最权威、最经典的文章来佐证你的研究。

(3)文章主体部分所涉及的基础理论和方法需要在引言中描述。

(4)引言中需要包含文章的主要结论。

(5)引言中所有第一次出现的专业术语需要使用全称而非缩写。

问题的实质即研究所要解决的核心,同样的核心问题在不同的条件下会有不同的外在展示,而引言首先要做的就是指出所要解决的核心问题是什么及所涉及的条件范围。简洁的研究背景描述可以很好地丰富研究对象的发生范围,研究意义的呈现则可以升华文章的层次,帮助文章不再陷入仅仅解决为什么而不解决怎么办的尴尬处境[3-5]。不仅如此,对研究背景和研究意义的描述,会极大地激发读者的阅读兴趣,吸引读者继续向下阅读。引用最相关、权威、经典的文献可以帮助读者对文章的研究课题进行快速分类,而和权威、经典的文献建立联系可以帮助非本专业的读者对文章涉及的研究内容有个初步认识,进而快速进入文章所提及的研究课题。不少读者可能会对第四点产生怀疑,毕竟科技论文的结论往往会作为一个独立的部分进行撰写,而在引言中不提及研究主要结论是不是会营造一种未知的氛围以让读者不断向下阅读而在结尾结论处大吃一惊呢?答案是否定的。不同于文学作品的写作,科技论文要求作者在开头处即给出本文的主要结论以帮助读者判断本文的阅读价值;同时,将研究的结论埋藏于文章的最后阶段可能会导致读者在未读到结论的时候就失去耐心。"Reading a scientific article isn't the same as reading a detective story. We want to know from the start that the butler did it."[6]第五点的目的是显而易见的,同样的缩写在不同行业中可能代表不同的意义,因此如要在文章中使用英文缩写,在第一次出现该专业词汇时必须使用这个词的英文全称,并用括号表明其英文缩写,便于后续文本表述使用。

4.3 引言的组织与结构

不同于论文的其他部分,引言所描述的内容较为多样,包含了研究的背景、意义、方法等,因此需要特定的结构进行组织。一般而言,引言可以进一步分为研究背景及意义、文献回顾、研究方法和重要结论四个部分[7,8]。

4.3.1 研究背景及意义

在引言的第一部分,作者需要介绍研究的背景环境,即在什么样的大背景下催生出怎样的问题,而文章核心问题的提出,往往是循序渐进的,是从较为宽泛的一般性事实中提炼而出的,而非直接在文章开头处给出。具体来说,在工程类科技论文的写作中,如新技术下催生的问题或者是对新现象下形成机理的探讨等这样的表述都可以作为研究背景进行开篇。

在描述完成研究的背景后,就需要对进行该项研究的意义进行表述。正如上文所说,研究意义的撰写是对研究主旨和实际应用意义的升华。任何科学研究都是为了现在或是未来解决某一具体的实际问题而进行的。因此,在引言的第一部分,作者需要写出研究的意义,以让读者知晓研究的应用领域和价值。

4.3.2 文献回顾

前人的研究是自身研究的理论基础,也是占据引言最大篇幅的部分,文献回顾既要联系自身研究和其他文献的逻辑关系也要串联起不同文献之间的关系,如递进关系、补充关系等。文献回顾的引文篇数从几篇到几十篇不等,具体应根据文章的类别和领域进行判断。

比如，文章研究了其他学者未研究的问题，则需要回顾其他学者先前撰写的大量文献，而若是仅仅探讨一个刚刚提出的科学前沿的小问题，则只需引用与此问题相关的文献即可。总之，文献回顾的长度应根据自身研究的领域来决定，务必做到简洁凝练、通俗易懂地完整介绍出关于此问题的研究进展和与文章的逻辑关系。根据所回顾的文献可进一步提出研究的问题，问题的提出可以用以下几种方式[9]。

(1) 其他学者已经研究过，但是未研究得十分透彻或不是十分完善的课题。
(2) 为解决某些问题而进行的研究中出现的新问题。
(3) 过去的研究在新兴的行业和领域中的应用或者改进与应用。
(4) 为解决过去的两种对立冲突的理论而进行的研究。

4.3.3 研究方法

研究方法的介绍既包括对试验性研究中实验方法的介绍，也包含理论性研究中理论模型的选取。在研究方法的介绍中，不仅要保证其方法本身介绍得具体、细致，也要对研究方法选取的缘由进行解释，最好能与文献回顾中其他学者所提出的方法进行比较。如果所涉及的研究方法是研究的主要创新成果，则可以在引言中先与其他研究方法进行对比，介绍其优点再在正文主体部分对研究方法进行更加详细的介绍。

4.3.4 重要结论

如上文介绍引言写作的基本原则中所描述的那样，重要结论需要在文章的引言中进行介绍，其理由也已详细说明，这里不再赘述。重要结论应区别于一般结论，一般结论可在文章的结论处单独详细撰写，而在引言的最后部分介绍的重要结论则是对研究核心成果的粗略展示，以起到让读者更快了解研究成果的作用。

4.4 引言的写作方法与技巧

引言的篇幅往往较为庞大，很多科研新手在第一次撰写引言的时候都会产生困惑。接下来，会从引言的时态选择和回顾文献的选取与介绍这两个方面解析引言的写作技巧。

4.4.1 引言的时态选择

有人说，科学研究的成果都是客观事实，因此，在引言的写作中应该使用一般现在时。这句话只说对了一半，在引言的写作中，一般现在时的应用的确占据了引言的主要部分，但是其他时态也有特定的应用情景。时态的运用规则可归类如下[10,11]。

(1) 对于已经广受认可的行业高被引、经典论文中所描述的研究结论及一些普遍的自然规律的引用使用一般现在时。例如，According to the experiment carried out by Max

Planck,the value of Plank constant is approximate 6.626 068 96×10^{-34}.

（2）当引用其他学者进行的研究成果、介绍其他研究开展的背景时，句子使用一般过去时，而描述研究结果的句子需要根据情况使用一般过去式或一般现在时。在（1）的例子中，由马克思·普朗克进行的实验由于发生在过去所以使用一般过去时，而普朗克常量已经被作为一个普遍规律进行使用，所以用一般现在时。又比如，Zhang found that the increase of the use of loops reduced the efficiency of computation in that test. 该定语从句的时态选择了一般过去式，是因为句子中描述的现象还没有作为普遍的事实。

（3）在描述领域中最近发生且对现有研究工作产生一定影响的工作时，应该使用现在完成时。如 Computer technology has been widely applied in civil engineering。

（4）在提出研究课题时，如果研究问题是一个普遍的现象，则用一般现在时。例如，The work of X remains a question. 但是如果是介绍一个其他学者的研究中一直没有解决的问题则需要使用现在完成时。比如，Although lots of studies have been done on X,few attention has been paid to Y。

（5）在介绍文章重要结论时，如果所介绍的研究结果是不受时间影响的普遍现象，则可以使用一般现在时。例如，In this research,a new analytical methods is introduced to describe the effect of X. 如果所介绍的研究结果是局限于研究过程中所进行的实验或是过去的某种特定环境时，由于实验发生在过去或者特定的环境已经过去，应使用一般过去式。例如，In the last experiment,the formation of X was investigated. 又如，The image of Halley's Comet was lastly captured through New Technology Telescope on January 10th,1994.

4.4.2 回顾文献的选取与介绍

回顾文献是串联起文章研究与其他学者研究的关系，并说明文章研究的创新性和实用性。在选取与介绍其他学者的文献工作中，一般使用以下一些技巧[12]。

（1）所选回顾文献的类型需要根据文章自身的研究主题进行确定。如果是根据其他学者现有研究进行的拓展研究，则最好从研究课题的经典文献开始介绍，逐步介绍到与文章研究课题所拓展的学者研究。如果所研究的问题是某领域新发现的问题，在前人研究较少的情况下可以结合其他领域的相似工作进行回顾，如问题的研究已经有所积淀则可以仅仅引用这些文献即可。当然，无论是新课题还是拓展研究，如果所研究的问题是位于一个大环境下的子问题，作者则应当在研究背景中对大环境进行介绍，在文献回顾中只引用关于研究问题的文献。

（2）在介绍其他学者的研究工作时，只需要做到介绍学者的姓名和工作，不需要细致入微地介绍其工作的开展情况。如果是文章研究的拓展基础，则应该联系文章说明其研究的优势与不足。

（3）回顾文献涉及不同的领域时应该分开介绍，每个领域内的文献应该按照时间顺序依次回顾。

4.5 实例分析

现通过笔者的三篇英文引言实例,说明英文引言的写作方法。

实例一 New method to calculate the apparent phase velocity of open-ended pipe pile[13]

  Pipe pile is widely used all over the world for its apparent advantages such as good adaptability, high bearing capacity, and remarkable economic results. However, various defects may occur during pipe pile installation regardless of the driving method. Low strain integrity testing is a cost-efficient and straightforward method for evaluating pile integrity (Likins and Rausche, 2000). For open-ended pipe piles, the interpretation of the measured signal faces more challenges due to the existence of soil plug, which affects not only the reflected signal waveform but also the apparent phase velocity of pipe piles(Guo and Ke, 2011). Finding a rational theoretical model to simulate the dynamic interaction between the soil plug and pipe pile during low strain testing, is of great importance in improving the testing accuracy and reducing miscalculations or misjudgments.

引言第一段首先介绍了研究的背景环境,阐明了土塞效应对开口管桩动力响应和视波速具有重要影响,表明需要合理的理论模型对其影响机理进行探究——体现出研究必要性和研究需求。

  The "pile within a pile" model pioneered by Heerema and De Jong(1979), which respectively discretizes the soil plug and the pipe pile volume into lumped nodes and springs, with frictional forces between the corresponding soil nodes and pile nodes, is often used in pile driving analysis, and estimating the static soil resistance distribution using the dynamic testing method(Randolph and Simons, 1986; Matsumoto and Takei, 1991; Liyanapathirana et al., 1998, 2001). However, this model is rarely utilized in the low strain testing of pipe piles, mainly because that the "pile within a pile" model is based on the one dimensional stress wave theory(Smith, 1960), whereas large diameter pipe piles exhibit serious three-dimensional characteristics during low strain test(Liao and Roesset, 1997; Chow et al., 2003; Chai et al., 2010; Ding et al., 2011; Zheng et al., 2016a, 2016b, 2017; Li et al., 2017; Li and Gao 2019). The measured

引言第二段阐述了已有的土塞与管桩相互作用模型,即"桩中桩"模型和平面应变模型,分析了这两种模型的不足,为该文模型的提出做好准备——体现出现有研究成果的不足。

reflected signal of low strain testing of pipe piles may be misinterpreted using the one-dimensional stress wave theory. The plane strain model(Novak et al., 1978) and its simplified dynamic Winkler model(Lee et al., 1988; Wang et al., 2010) are two primary models to simulate the surrounding soil-pile interaction during low strain test. However, these two models cannot directly extrapolate to the soil plug-pipe pile interaction owing to the different boundary conditions, where the plane strain model assumes the surrounding soil to be an infinite medium, but indeed the soil plug is encircled inside the pipe pile, and the vibration energy cannot be propagated to the far-field(Randolph, 2003). Another apparent deficiency of the two commonly used models is that the effect of soil plug mass is not taken into consideration, which in fact remarkably affects the energy transmission and the apparent phase velocity of pipe piles.

The apparent phase velocity of pipe piles, which plays an essential role in length estimation and defect detection, is conventionally taken as the material longitudinal wave velocity measured before the pile installation. However, the apparent phase velocity cannot be treated as a constant when considering the soil damping effect. Makris and Gazetas(1993) modeled the pile surrounding soil as dynamic Winkler medium and reported that the pile phase velocity varies with frequency. Wu et al.(2017a) observed that the apparent phase velocity for the pipe pile segment with the soil plug decreases by nearly 50%. Nevertheless, none of the existing soil plug models can reflect this phenomenon. Therefore, a new soil plug model, namely, the additional mass model was proposed by Wu et al.(2017a, 2017b) to solve this problem. In this model, the soil plug is divided into small segments connected to the inner pile wall through the distributed Voigt model. Based on the additional mass model, Wu et al.(2017b) derived an analytical solution of the vertical dynamic velocity response at pipe pile head which matches well with the measured curve, and Liu et al. (2017) obtained an analytical solution of the torsional dynamic

引言第三段分析了土塞视波速的影响因素，基于此提出附加质量模型，并对附加质量模型的合理性和初步应用进行了介绍——体现该文研究的创新性。

response at pipe pile head. Furthermore, Liu et al. (2018a) provided a more accurate solution with the consideration of stress wave propagation in both the vertical and circumferential directions, and proposed the double-velocity symmetrical superposition method to eliminate the high-frequency interference at pipe pile head without increasing the predominant period of impact pulse(Liu et al. 2018b). The merits and accuracy of the additional mass model have been confirmed through a series of model and field tests, but, still, two factors limit its application: (a) the apparent phase velocity is back analyzed from the dynamic velocity response, which is not straightforward and needs sophisticated derivation; (b) the selection of its parameter combination has not been clarified.

In light of this, the objective of this paper is to present an analytical solution to calculate the apparent phase velocity of the pipe pile segment with soil plug(APVPSP) based on the additional mass model. A parameter combination suitable for most commonly used concrete pipe piles is recommended. The rationality and accuracy of the developed solution have been confirmed through the comparison with the solution using the soil-plug Winkler model and experimental results. A parametric study is also conducted to investigate the influence of soil plug, pipe pile and incident impulse on the APVPSP.

引言第四段介绍该文的研究目标和研究方法与过程——展示该文的结构与成果。

实例二 DEM analysis of the sand plug behavior during the installation process of open-ended pile[14]

Open-ended piles, with the merits of low installation resistance and high loading capacity, are widely used in the construction of onshore and offshore structure foundations. During the installation process, soil mass underneath will be pushed into the pile, giving birth to the so-called "soil plug". Depending on the relative displacement between the pile and the soil plug, the open-ended pile may penetrate as plugged, partially plugged and unplugged. At initial penetration,

引言第一段首先介绍了研究的背景环境,阐明土塞效应是判别开口管桩为闭塞、部分闭塞和不闭塞的重要指标,并指出土塞是理解开口管桩承载性能和沉桩过程的关键因素——体现出研究必要性和研究需求。

the height of the soil plug is almost equal to the penetration depth; with increasing penetration and accumulation of inner shaft resistance, the soil mass inside the pile may act like a perfect plug, preventing or partially restrict soil flowing into the pile. At this stage, the open-ended pile may assume the installation characteristics of a closed-ended pile. Different modes of soil plug behavior may highly affect the installation resistance and bearing capacity of the open-ended pile, making it necessary to understand the plug behavior during the pile installation process.

A large amount of valuable work has been devoted to research on the plugging effect of open-ended piles. The notions of the plug length ratio(PLR) and the incremental filling ratio(IFR) have been proposed to quantify the degree of soil plugging. The PLR is defined as the soil plug length per penetration depth, while the IFR is defined as the increment of soil plug length per increment of pile penetration. Some scholars believed that IFR is a better indicator of the plugging effect than PLR. However, the complexity of plug behavior has made the above two parameters hard to obtain and explain since the plugging effect may be highly influenced by the installation method, pile geometry, soil conditions, etc. Kishda and Isemoto performed a series of push-up load tests, where sand plugs inside steel pipe piles were pushed up using a rigid platten. Test results indicated the load was mainly supported by the shaft resistance at the bottom two pile diameters of the soil plug and the values of earth pressure coefficient increased gradually from the top to the bottom of the soil plug. Based on similar tests, Hight proposed the idea of critical height for soil plug; when soil plug height is larger than the critical value, the soil mass could lock up like a perfect plug. He also indicated that the increase of soil density and decrease of pile diameter may lead to the decrease of Hcri value. O'Neill and Raines found that during static loading, a very high load transfer occurred over the bottom 3-4 D of the sand plug, where D is the inside diameter of the pile. An arching theory was later introduced by Paikowsky to explain the load transfer mechanism inside

引言第二段阐述了开口管桩沉桩过程的已有成果，包括土塞增长率、土塞长度比、土塞位移场和应力场等，并分析了现有成果不完善的地方——体现出现有研究成果的不足。

the soil plug. In terms of the installation method of open-ended pile, research on driven piles obviously outnumbers that on jacked piles. Based on the centrifuge test on model piles, de Nicola and Randolph observed that during pile driving, the plug length tend to grow with increasing relative density, while the trend was opposite during jacking; it also suggested that the jacked pile was more likely to plug than the driven pile. Also, a growing body of evidence from field and laboratory tests has revealed that most open-ended pile would keep the 'unplugged' or 'partially plugged' state during driving, while it may plug under static loading. Other factors like in situ stress, relative density, pile diameter and penetration depth have also been studied. Apart from experimental study, Randolph et al. proposed a simple one-dimensional model to calculate the stress distribution in a partially drained sand plug and the idea of "active plug length" was introduced, analytical and numerical results revealed that the vertical effective stress varied almost logarithmically along the "active plug". Numerical models based on the finite element method were constructed by Liyanapathirana et al. to explore the driven response of infinitely-long and thin-walled open-ended piles. Attention has also been paid to the dynamic response of soil plug from both experimental and analytical analysis.

The above and related research has greatly advanced our understanding about the deformation and capacity characteristics of soil-pile system, however, considering the difficulty to measure the stress and strain state in field and laboratory tests and limitations of the Finite Element Method(FEM) to analyze large deformation and discontinuous problems, the Discrete Element Method(DEM) presents a satisfactory alternative to investigate both the macroscopic and microscopic behavior of granular soil. Based on DEM, Guerrero analyzed the influence of end shape of driven piles on particle crushing and driving resistance, where the highest resistance and crushing amounts were observed for flat-ended piles than open-ended piles and triangle-tip piles.

引言第三段对比了现有研究管桩成桩过程方法 FEM 和 DEM 的优缺点，指出 DEM 在管桩沉桩过程分析中的优势，并将 DEM 定为该文的研究工具——体现该文研究的创新性。

Zhou et al. conducted model tests and DEM simulations on open-ended piles jacking into sands, the formation process of soil plug was successfully reproduced. Thongmunee et al. carried out experimental study and DEM simulation of push-up load tests on dry sandy plugs, with the focus on the influences of packing state and aspect ratio of the sand plug on the bearing capacity. Zhang and Wang simulated the monotonic jacking process in a 3D centrifuge model test; a microscopic view was provided to reveal mechanism considering stress distribution, stress path, particle movements and contact force mobilization. Esposito performed a sensitivity analysis of single pile installation. Numerical results indicated that installation methods and particle rotation would have a great influence on pile performance, while the influence from penetration velocity and pile-soil friction was less. DEM analysis has gained wide acceptance in pile-related studies, however, the majority of the research has focused on the macroscopic perspective for close-ended piles or CPT tests, while microscopic investigation on the penetration characteristics of open-ended piles has been less favored.

Based on the above analysis, this study aims to construct a DEM model to analyze the plug behavior during the jacking and subsequent loading process of open-ended pile. The commercial DEM software Particle Flow Code in Two Dimensions(PFC 2D) is adopted to reveal the macroscopic and microscopic responses of the soil plug. The remainder of this paper is outlined as follows: first, a homogenous sand assembly is generated and simulation of the installation process is introduced. Second, numerical results are provided in detail to reveal the macroscopic and microscopic soil response, based on the distribution of porosity and stress state, particle movement and the rotation of principal stress. At last, discussions of the DEM modelling results are provided.

引言第四段介绍该文的研究目标和研究方法与过程——展示该文的结构与成果。

第 4 章　引言与研究背景

实例三：**The impact of ground soil properties on the structural integrity of high-fill reinforced concrete culverts**[15]

High-fill reinforced concrete (HRC) culverts have been widely used as substructures for water, vehicle and pedestrian conveyance. The results of a random field survey of 102 HRC culverts, which were located on highways in 5 provinces (Hunan, Hubei, Sichuan, Shanxi, and Henan) in China, demonstrated that structural problems occur frequently in culverts. Typical characteristics of damaged HRC culverts are summarized in Table 1, which shows that the ratio of damaged culverts to total surveyed culverts is 59.8%. In particular, the damage ratios of slab culverts, arch culverts, pipe culverts and box culverts are 64.8%, 54.8%, 63.6% and 33.3%, respectively. Significant damage characteristics include structural cracks and differential settlements of culvert foundations. In this survey, the damaged culvert is defined as the crack width on the culvert is more than 0.2mm or the differential foundation settlement is more than 30mm based on the *Chinese Code for Design of Highway Reinforced Concrete and Prestressed Concrete Bridges and Culverts*.

〔引言第一段首先介绍了研究的背景环境，阐明了高填充混凝土涵洞具有广阔的应用前景，并指出该涵洞会存在结构裂缝和不均匀沉降的风险，亟需深入研究——体现出研究必要性和研究需求。〕

These problems occur in HRC culverts for two important reasons: ① vertical earth pressure on culverts and ② impact of ground soil properties on culverts. Compression of the backfill mass at both sides of the culvert is larger than that of the RC culvert due to the differential stiffness. Thus, vertical earth pressure concentrates on the top of RC culverts. The vertical earth pressure concentration on culverts will be more significant for a larger differential stiffness between a culvert and its adjacent backfill.

〔引言第二段指出涵洞出现这些问题的原因：涵洞上的垂直土压力和地面土体特性的影响，指出该文拟关注的两个问题——明确该文的研究思路。〕

Marston pioneered the research concerning vertical earth pressure on culverts based on analytical and experimental methods. Following the research of Marston, Spangler analyzed the vertical earth pressure on rigid pipe culverts and discussed the key factors that influence the load on underground conduits. Karinski et al. investigated vertical earth pressure on rigid

〔引言第三段阐述现有涵洞的研究成果，指出现有成果缺少关于地面土体性质对涵洞结构完整性和应力状态的研究——体现出现有研究成果的不足。〕

culverts under fill and traffic loads through theoretical analysis and discussed the influence of structural deformation and dimension on vertical earth pressure. Kim et al. examined vertical earth pressure on rectangular culverts under three different installation conditions (embankment installation, trench installation and imperfect trench installation) by numerical simulation. Bennett et al. studied the structure stress states and vertical earth pressure on RC box culverts by conducting field tests and found that the height of the backfill over the culvert was the main factor that influenced vertical earth pressure and internal structural forces. Kang et al. analyzed the soil-culvert interaction under two conditions, i. e. , imperfect trench installation and embankment installation, using numerical simulation. Valsangkar et al. investigated the backfill-culvert interaction and detected the earth pressure acting on culvert under embankment installation and induced trench installation conditions, using field and centrifuge test and numerical simulation. These studies thoroughly addressed the subject of vertical earth pressure on culverts and the relationship between fill load and structural stress states. However, few studies have focused on the impact of ground soil properties on the structural integrity and stress states of HRC culverts.

When investigating the structural integrity and stress states of HRC culverts, the mechanicalproperties of the subgrade layer cannot be disregarded. The objective of this study is to discuss how the following two principal issues affect HRC culvert stress states and integrity: ①the ground bearing capacity of HRC culverts and ②ground treatment of HRC culverts.

> 引言第四段介绍该文的研究目标——展示该文拟可能取得的成果。

  由上文所举的三个例子来看,引言的写作思路如下:研究背景与意义—文献回顾—方法介绍—主要成果。各部分之间互相依存,紧密结合,其目的是将所研究项目的故事徐徐展开。国内材料研究学者在撰写英文研究论文时,应当谙熟英文学术论文引言的体裁结构与表达方式,遵循英文引言写作规范,提高英文论文引言的写作质量与学术水平。建议每个工作日阅读一两篇英文为母语作者的专业论文,并建立语料库,注重积累,对提高英语论文的写作水平会有事半功倍的效果。

本章参考文献

[1] 贺萍.英文科技论文中引言的撰写与编辑[J].中国科技期刊研究,2005,16(2):261-263.

[2] 张美慧.英文科技期刊引言写作的常见问题探析[J].创作空间,2016(2):68-71.

[3] 李娜,陈欣.英语学术论文引言部分的跨学科体裁分析[J].科研园地,2018(5):88-93.

[4] 陈菁,于学玲,史志祥.中外生物类英文SCI收录期刊论文引言部分的体裁分析与对比[J].中国科技期刊研究,2019,30(2):143-148.

[5] 蒋婷,徐娟.英文法律类论文中引言的体裁研究[J].社会科学研究,2013(2):203-208.

[6] DAY R A, GASTEL B. How to write and publish a scientific paper[M]. 6th ed. New York: Greenwood Press, 2006.

[7] 雍文明.英文医学论文引言结构与编写模板研究[J].中国科技期刊研究,2018,29(4):362-367.

[8] 杨建新,李平艳.英汉学术论文引言中引用的修辞劝说对比[J].外语与翻译,2018(2):38-45.

[9] 刘锋,张京鱼.农业科技期刊英文论文引言结构与内容特征及写编建议[J].中国科技期刊研究,2017,28(12):1134-1140.

[10] 许耀元,王磊.国际矿业英文学术期刊中外作者论文引言的语境功能标记性主位对比分析[J].河南机电高等专科学校学报,2010,18(6):80-83.

[11] 陈竹,李洁,王华菊,等.英文材料研究论文引言的写作[J].编辑学报,2017,29(S1):40-43.

[12] 丹阳,刘俭,彭勇,等.国际高质量科技论文引言逻辑模型研究[J].山西科技,2019,34(1):142-146.

[13] WU W B, LIU H, YANG X Y, et al. New method to calculate the apparent phase velocity of open-ended pipe pile[J]. Canadian Geotechnical Journal, 2020, 57(1): 127-138.

[14] LI L C, WU W B, EL NAGGAR M H, et al. DEM analysis of the sand plug behavior during the installation process of open-endedpile[J]. Computers and Geotechnics, 2019, 109: 23-33.

[15] CHEN B G, SUN L. The impact of ground soil properties on the structural integrity of high-fill reinforced concrete culverts[J]. Computers and Geotechnics, 2013, 52: 46-53.

第5章　正文其他部分的基本要求与写作要点

5.1　概　述

正文是科技论文的主体部分，通常包括引言(introduction)、材料与方法(materials and methods)、研究结果(results)、讨论(discussion)、结论(conclusion)等部分，每一部分有各自的结构和写作特点。第4章介绍了引言的写作方法，本章将依次对正文的研究对象与方法、研究结果、讨论、结论部分的基本要求和写作要点进行介绍。

5.2　研究方法与内容

5.2.1　基本要求

对科技论文来说，研究方法与内容这一部分的主要目的是将一个科学问题的研究过程转化为具体的操作方法，并提供所选方法的具体实施细节以便相关领域的读者在阅读完后能对作者的研究成果进行重复。因此，在这一部分的写作过程中要注意结果的可重复性、方法的可靠性及前后的逻辑性。

研究方法与内容部分主要包括对研究对象和研究方法的描述，对于工程专业的学生来说，研究对象多依托于一定的工程问题，研究方法主要包括试验方法、理论方法和数值模拟方法等。

5.2.2　写作要点

1. 篇章结构及写作方法

研究方法与内容部分没有一定的写作模板，根据文章需要或采用方法的不同可以按操作步骤层层深入，也可以按章节内容另设分层副标题。考虑到相关领域的研究者对研究方法与内容部分较为熟悉，这一部分也是较为容易被读者略过的。因此，建议这一段采用主标题加副标题的结构，使整段结构更加清晰，也使读者能更加快速和准确地定位到感兴趣的部分。

在写作内容上,要把研究对象、实验(或操作)方法、实验器材、详细步骤等如实写出,并提供理论依据,以便读者借助相同的仪器或材料可以得出类似的结果。在写作方法上,要注意以下三点:可重复性,即把所选方法的每个环节、步骤、要点都叙述清楚;具体性,即对研究对象的物理力学性质和研究方法的细节描述要具体,使之具有可操作性;描述性,即对材料的特点及所采取方法的实现步骤等进行描述时,可采用流程图、表格、照片等方法[1,2]。

以下结合工程科技论文中常见的几种研究方法,给出具有代表性的例子。

例1 现场试验类:*Plugging effect of open-ended piles in sandy soil*[3]

2 Methods available for plugging effect on open-ended piles

3 Field load test

  3.1 Site investigation

  3.2 Installed test piles

  3.3 Dynamic load test

  3.4 Static load test

例2 室内实验类:*Comparing the slaking of clay-bearing rocks under laboratory conditions to slaking under natural climatic conditions*[4]

2 Methodology

  2.1 Sampling

  2.2 Laboratory testing

  2.3 Exposing clay-bearing rocks to natural climatic conditions

  2.4 Quantifying the amount of slaking

例3 数值模拟类:*DEM analysis of "soil"-arching within geogrid-reinforced and unreinforced pile-supported embankments*[5]

2 Numerical model

  2.1 Description of selected cases

  2.2 Simulation procedure and parameter determination

    2.2.1 Sample preparation method

    2.2.2 Embankment fill properties

    2.2.3 Subsoil properties

    2.2.4 Geogrid properties

  2.3 Simulation model for the selected cases

  2.4 Validation of the simulated model

例4 理论分析类:*A new interaction model for the vertical dynamic response of pipe piles considering soil plug effect*[6]

2 Analytical model

  2.1 Conceptual model and assumptions

  2.2 Equations of motion

    2.2.1 Governing equation of soil

2.2.2 Governing equation of pipe pile
2.3 Boundary and initial conditions of soil-pile system
2.3.1 Boundary conditions of pipe pile
2.3.2 Initial conditions of soil-pile system

可以看出，在上述实例中，各论文的结构安排与所选方法不尽相同，但都很好地对所研究对象及所采用的研究方法进行了整体的介绍。

此外，采用副标题的结构安排能使文章读起来更为流畅，也能更好地让读者迅速定位到他/她所关注的部分。

2. 对研究对象的描述应清楚、准确

对研究对象的描述，不同的学科有不同的要求，这里很难加以详细的展开。从总体上来说，要注意说明研究对象选择的必要性，也就是对为什么选择这种研究对象最好有一定的说明。同时，要对研究对象的基本物理力学参数进行介绍。对于实验类研究，要给出样本来源，例如采样地点、数目、种类。当样本较多时，可通过表格的形式对样本进行整理与展示。对于理论类或数值模拟类研究，要分别给出实际参数与模拟参数，同时对参数选取的合理性进行验证。

3. 对研究方法的描述要详略得当、重点突出

研究方法介绍研究是如何开展和实施的，通常按照研究开展的时间顺序进行介绍。由于研究领域和采用方法的不同，其内容也会有所差异，一般包括参数采集方法、试样制备方法、参数分析方法、模型制作、程序编制、建模分析等。如果没有较为明确的时间顺序，也可以按照研究开展的逻辑顺序或重要性程度描述研究步骤。当某一研究方法步骤较多，此时单纯用文字进行描述显得较为杂乱，建议附加流程图加以说明。流程图的形式可以多种多样，可以是文字式的，也可以是文字和示意图相结合的，只要能将研究方法的具体步骤阐述清楚即可(图5-1)。

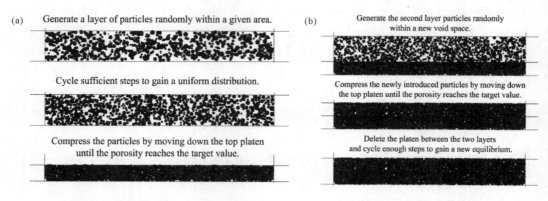

图5-1 流程图示例——Flow chart for generating assembly by the use of IMCM
(a)the first layer；(b)the second layer[5]

描述研究方法应遵循的原则是，给出关键信息以便让同行能够重复实验，避免引入关于结果或发现方面的内容。同时,对方法的描述又不应太过细致,以致占据过多篇幅,导致读者难以得出有效信息。对方法的描述应注意以下几点:

(1)若运用标准规定的方法或众所周知的方法,不必花过多篇幅介绍,只需写明采用什么方法即可。

(2)若是引用别人报道过的方法,则引用相关文献,并简单介绍方法的操作步骤。若这种方法大家并不熟悉,或是采用该方法的文献影响力较小,可稍加详细描述。若对别人的方法有实质性改进,应重点写明改进的部分,并且说明理由。

(3)若是提出新的方法,且未曾发表,则要详细介绍,以便他人重复。

5.2.3 语态与时态的运用

在研究内容与方法部分中,尽量避免采用第一人称,可以在一定程度上使用主动语态,但大部分内容建议使用第三人称和被动语态。同时,本节中大部分内容应该以过去时来描述[1,2],具体可参考以下几点。

(1)若描述的内容为不受时间影响的事实,采用一般现在时。如"The plugging effect can be quantified by using the plug length ratio(PLR)and the IFR."[7]。

(2)若描述的内容为特定、过去的行为或事件,则采用过去式。如"Field load tests were conducted at the Kwangyang plant construction site to investigate the plugging effect of open-ended piles in sandy soil."[7]。

(3)方法部分的焦点在于描述所采用方法的主要步骤及所研究的对象,由于所涉及的行为与研究对象是讨论的焦点,而且读者已知道这些行为的执行者就是作者本人,因而一般都习惯采用被动语态。如"All piles were driven with a DKH-13 hydraulic hammer (130kN)."[7]。

(4)如果涉及表达作者的观点或目的,则应采用主动语态或不定式结构。如"The gap between the outer and inner pile was welded to prevent any intrusion of soil during the tests."[7]。

除了上述几点需要注意外,研究方法与内容部分同时需要准确和详细的语言描述,因为这一部分是介绍具体的研究对象和研究方法。在语言组织过程中,应尽量避免使用模糊的描述,如 sometimes、on occasion、maybe、approximately 等。

5.3 研究结果

5.3.1 基本要求

结果部分的作用是采用适当的文字或图表形式,将整理和归纳后的研究结果客观、有序地呈现给读者。论文的前面部分(引言、研究背景、研究方法)是为了解释为什么和怎么样获得这些结果,论文的后面部分(讨论与结论)则是进一步解释这些结果背后的意义和价值。因此,关于结果部分的描述可以看作是文章的中心和展开的基础。

需要注意的是,结果的核心内容应该是经过分析与归纳得出来的数据,而不是原始数据,更不是原始记录。所有的研究成果均要围绕研究主题有逻辑、有层次地展开,与主题无关的部分不宜全部列出。注意,研究结果应该是客观的(objective)、经过归纳的(summarized)、与研究结论有关的(conclusion-related)。

结果部分的主要内容包括:

(1)结果的介绍。即指出结果在哪些图表中列出。

(2)结果的描述。即描述重要的分析结果。

(3)对结果的评论。对结果的进一步说明与解释、已已记录的结果进行比较等。

5.3.2　写作要点

1. 篇章结构

由于研究领域和研究方法的不同,结果部分没有特定的结构要求。结果部分可以按照实验先后的顺序或结果的重要程度介绍,建议按研究结果间的逻辑顺序展开。对结果的描述要能让读者看出研究的目的是什么、逻辑结构是什么、结果分析前后的顺序如何。一个好的结果描述能让读者顺着作者的思维一步一步地获得答案。通过文字告诉读者本研究发现了什么,通过表格将具体的数据有条理地展现给读者,通过图形使读者能够直观地理解研究结果。

2. 信息的提炼

缺少论文写作经验与训练的人,很容易将结果部分写成实验报告,即将所有的研究结果一一堆积到论文中,以体现自己的大量工作,其实,这样往往会适得其反。科技论文的写作应该突出有科学意义和具有代表性的数据,而不是没完没了地重复一般性数据。因此,要学会高度概括和提炼关键信息。当然,对不同的研究结果应采取不同的写作方式。对特别重要的研究结果或者文章中的创新点应提供原始数据并对研究结果进行详细的介绍,对于一般性数据或通过常规实验得出的结果只需提供主要结果。

需要注意的是,若论文的讨论部分单独成章,对结果的详细讨论应留到该部分,但仍需对研究结果提供必要的说明,以便读者能清楚地了解本研究的主要研究结果及其重要性。

3. 副标题的使用

副标题的目的是使文章更有条理,帮助读者清楚地了解研究结果的主要信息,并找到他们感兴趣的内容。一般来说,副标题应尽量概括该部分的主要内容。可以把结果部分看作实验结果的集合,并尽可能清晰和简洁地总结在图表、表格、方程和示意图中。记住不要简单地罗列结果,要进行必要的分析,各副标题间要有一定的逻辑联系。

4. 图表的使用

一图胜千言(A picture is worth a thousand words.)。这句话对科技论文的写作同样适用。相比于文字,图表能更加直观地描述出结果中数据的特点及变化趋势。需要注意的是,凡是能用文字说明的问题,或是通过简单的文字即可表述清楚的结果特点,则尽量不要使用图表。当论文篇幅较长或数据较多且文字难以全面描述结果特征时,则需要利用图表对结果内容进行描述。若采用图表,文字部分则应分析图表中数据的重要特征或趋势。切忌仅对图表中的数据进行简单的重复描述,而忽视描述图表中数据的意义。同时,应避免使用冗长的词句来介绍或解释图表,避免将图表的序号或标题作为段落的主题句。

5. "坏"结果的介绍

在进行科学研究时,有时会得到一些与预期不同的结果,这可能意味着研究方法中的某一个环节出现了错误,或是研究对象的某些参数没有达到标准。这时,就需要研究者对得到的结果进行分析,找出与预期不符的原因。还有一种可能是,得到的结果揭示了一种新的现象。不管是哪种情况,得到的结果都具有重要的意义。不要误认为与预期不符的结果都不值得报道,或者认为它们是"坏"结果。如果严格遵循了研究方法,并对它进行详细的报道,那么得到的结果就是你的结果,只是它需要得到合适的解释。请记住,很多重大的发现最开始都来自"坏"结果。

5.3.3 语态与时态的运用

撰写结果部分,主动语态、被动语态都可以使用,以被动语态为主。在时态运用方面,论述普遍现象可以采用一般现在时,有时也用一般过去时,注意避免在同一段落中频繁地转换时态,具体可以参考以下几点。

(1)当描述某一结果呈现在哪些图表中时,常用一般现在时。例如,"Grain size distribution curves of slaked material for laboratory and roof specimens are shown in Fig. 6."[4]。

(2)叙述或总结研究结果的内容是关于过去的事实时,通常采用过去时。例如,"As the slake durability test progressed from the 1st to the 5th cycle, an increasing amount of slaking was observed for all four types of clay-bearing rocks."[4]。有时,在论文中也会用现在时描述结果。它与使用过去时的差别是:使用现在时说明该结果是在研究过程中所揭示的是普遍事实,使用一般过去时则说明本次研究结果是在某些特定情况下发现的事实。

(3)对研究结果进行说明或由它得出一般性推论时,多用现在时。例如,"Generally, the wide range of DR values exhibited by all four types of clay-bearing rocks over the five cycles of slake durability test(Fig. 4)reflects their geologic diversity."[4]。

(4)将本文研究结果与其他实验数据与理论模型之间进行比较时,多采一般现在时(这种比较关系一般为不受时间影响的逻辑上的事实)。例如,"The good agreement between the calculated and measured responses confirms that the additional mass model can simulate the dynamic interaction between the soil plug and pipe pile."[6]。

5.4 讨 论

5.4.1 基本要求

在对研究结果进行描述之后,作者需要在文章的理论框架下对所得结果进行解释与推断,即讨论这些结果说明了什么问题,是如何支持作者期望获得的科学结论的,是否提出了新的问题或观点等。一个好的讨论,能让读者顺着作者的思路,逐步地、有逻辑地对研究问题进行分析[8]。

论文讨论部分的撰写没有特殊的格式要求,可以单独成章(discussion),可以与结果部分合并(results and discussion),也可以与结论部分合并(discussion and conclusion)。因此,作者应根据文章的结构和相应期刊的要求进行合理安排。

讨论部分应基于研究结果,结合基础理论和已有研究,对研究结果受影响的因素进行分析,对意外发现做出解释、建议和设想。其主要内容应包括以下几个方面。

(1)简要说明已获得的主要结果,并结合研究的目的和假设,讨论得到的结果是否符合期望? 如果不符合,为什么?

(2)对比讨论本文发现中与已有同类报道中结论的异同,哪些文献支持本文的发现,哪些文献的结论与本文不同? 如果不同,讨论一下产生这些差异的原因?

(3)应指出本文存在的不足之处或者局限性;讨论局限性产生的原因,可能对结果产生的影响,要证实本文的结论还需要进行哪些研究等。

(4)指出本文研究结果的理论意义和应用价值,提出进一步的研究方向、展望、建议和设想。

5.4.2 写作要点

一篇好的论文不仅仅是研究数据的罗列,更是以结果部分为基础和线索进行分析和推理,并表达在结果部分所不能表达的更深层次内容。水平一般的讨论只是就事论事,只能根据自己的数据得出相应的结果。水平高的讨论会把自己的数据放在一个背景中,与已发表的数据,模型或者观点进行比较。讨论部分是使研究成果得到升华的部分,在撰写讨论时应注意以下方面。

1. 段落结构

一般来说,对于讨论部分的撰写没有标准的格式要求,对于写作水平较高的研究者来说,可以按自己的写作风格进行撰写。但对一般研究者来说,采用一定的叙事结构进行撰写,论文的应用性和逻辑性更强。James Hartley[9,10]对讨论部分的结构给出了一些建议:

Step 1:restate the findings and accomplishments.

Step 2:Evaluate how the results fit in with the previous findings—do they contradict, qualify,agree or go beyond them?

Step 3:List potential limitations to the study.

Step 4:Offer an interpretation/explanation of these results and ward off counter-claims.

Step 5:State the implications and recommend further research.

在撰写讨论部分时,往往会讨论多个话题,此时可以将此部分分为若干子段落,段落之间可以是并列关系或者递进关系,但是要保证每一段都有一个主题,即每一段讨论一个主要话题,且每个子段落仍可遵循上述结构。

2. 源于结果,高于结果

讨论是根据研究的目的和结果所做的总结性、拔高性描述,主要是对研究所反映的问题进行分析和评价。有些作者在讨论部分仍喜欢强调引言部分已经明确的研究目的,并引用结果部分已经列出的具体数据,使讨论和引言、结果部分内容重复。这种做法既浪费篇幅,又容易使读者产生繁复之感,使讨论部分的作用得不到展现。

在讨论中,需要特别指出的是要保持与结果的一致性,也就是结果和讨论要前后呼应,相互衬托才可以。千万不要出现按照讨论的内容推出与结果部分相反结论这种事情,那就证明你的讨论思路是彻底失败的或者是你的实验过程出了问题。

同样需要注意的是,讨论的内容应当从实验和观察结果出发,实事求是,切不可主观推测,超越数据所能达到的范围,要慎用"首创""领先"等词组。如果把数据外推到一个更大的、不恰当的结论,不仅无益于提高作者的科学贡献,甚至已有数据所支持的结论也会受到怀疑。

3. 抓住重点,多角度分析

很多作者认为讨论部分是文章中最难写的部分,因为这一部分往往能体现作者研究问题的深度和广度。深度就是论文对于提出问题的研究到了一个什么样的程度,广度就是是否能够从多个角度来分析与解释所得的结果。

一方面,很多作者在进行讨论时,总希望面面俱到,这样便导致了讨论中多是一般性和概括性的分析,而抓不住重点。选择合适的结果在讨论部分中进行深入分析,是写好该部分首先面临的一个问题。一般来说,可以根据一个简单的原则来判断:如果你得到的结果体现了研究的独特性或创新性,那么这个结果就是应该要重点讨论的问题。有些结果和前人的研究相一致的,或者没有显著性的差异,那么就应该一笔带过而不要深入讨论;否则,那只是重复别人的工作而已,没有任何价值。讨论的一个重要作用就是要突出自己研究的创新性,体现出自己显著区别于他人的特点,区别的大和小是另外一个问题,重要的在于有区别。

另一方面,一些作者面对新颖的研究结果却不知道如何展开讨论,对于这些作者,建议对选中的问题按照一定的层次从多个角度来进行讨论。首先与已报道的类似结果进行对比,说明自己结论的独特性。其次要系统地阐述为什么会有这样的结果,方法可以有多种:从实验设计的角度、从理论原理的角度、从分析方法的角度,或者借鉴别人分析的方法,等等。无论问题大小、重要与否,都可以尝试从多个角度展开讨论。

4. 局限性与不足

任何研究,哪怕是发表在顶级期刊上的研究,也不可能毫无瑕疵,总会在某些方面存在一定的局限性。有些作者常常花很大篇幅强调研究的重要性和价值,但在研究的局限性方面很少提及,或者干脆只字不提。刻意地隐藏文章的漏洞,觉得读者看不出来,是非常不明智的。

事实上给出文章的局限性与不足恰恰是保护自己文章的重要手段。但是,在指出这些不足之后,随后一定要再一次加强本文研究的重要性及可能采取的手段来解决这些不足,为别人或者自己的下一步研究打下伏笔。通过给出不足与解决办法,把审稿人想到的问题提前给一个交代,同时表明你已经在思考这些问题,但是由于文章长度,试验进度或者试验手段的制约,文章中暂时不能回答这些问题。但是,通过你给出的一些建议,在将来的研究中解决这些问题是有可能实现的。

5.4.3 语态和时态的运用

在讨论部分,作者要对研究结果和发现进行分析、推断、演绎和推理,就要求作者具有很强逻辑思维能力和英语文字组织能力。这一部分,主动语态、被动语态都可以使用,但以主动语态为主。同时,应注意用语的准确性和简洁性,避免使用类似"Future studies are needed."这类没有实际作用的句子。此外,这部分时态比较复杂,要分清实验过程和结果(过去时)与分析意见(确定:现在时;不确定或假设:过去时)的区别、他人研究结果(过去时或现在完成时)与本研究结果(过去时)的区别、普遍适用的结论(现在时)与只适用本研究的结论(过去时)的区别等。同时,可采用适当的词汇来区分推测与事实。例如,选用 imply、suggest 等表示推测;或者选用情态动词 will、should、can、may、could、probably、possibly 等来表示论点的确定性程度。具体可参考以下几点:

(1)回顾研究目的与主要成果时,通常使用过去时。如"In this study, an artificial method was used to distinguish the non-scale range in the double logarithmic curve."[11]。

(2)如果作者认为所讨论结果的有效性只是针对本次特定的研究,需用过去时;相反,如果具有普遍的意义,则用现在时。如"For instance, it was observed that the developed soil arching structure would be transformed from the 'full arching' to the 'partial arching' with the increase in a(as shown in Fig. 5(d)—(f))… In other words, H_{crit} is correlated with the pile(or pile cap)width a for a given$(s-a)$."[12]。

(3)阐述由结果得出的推论时,通常使用现在时。使用现在时的理由是因为得出的是具普遍有效的结论或推论(而不只是在讨论自己的研究结果),并且结果与结论或推论之间的逻辑关系为不受时间影响的事实。如"Based on the observed deformation behaviours of the modelled pile-supported embankments, soil arching structures can be divided into three groups…"[12]。

5.5 结　论

5.5.1 基本要求

当对所有的研究结果进行分析与讨论之后，作者需要思考本研究中最重要的、总结性的结论，并呈现给读者，这就是结论部分。

论文的结论部分应反映论文中通过实验和理论分析后得到的学术见解。换句话说，结论应是整篇论文的结局，而不是某一局部问题或某一分支问题的结论，也不是正文中各段小结的简单重复。结论应当体现作者更深层的认识，且是从全篇论文的全部材料出发，经过推理、判断、归纳等逻辑分析过程而得到的新的学术总观念、总见解。

结论部分是作者发表观点和见解，给读者的精髓部分，应注意以下几点。

(1) 归纳性说明研究结果或发现。
(2) 结论性说明结果的可能原因、机理或意义。
(3) 前瞻性说明未解决的问题。

5.5.2 写作要点

1. 段落结构

结论部分可以单独成段，也可以与讨论部分合并为"discussion and conclusion"。如果文章篇幅较长，且结论结构清晰、条理分明，建议单独设置结论部分，可以起到鲜明突出的作用。结论部分的内容安排主要有两种结构，一种像法律条文一样，按顺序1,2,3,…列出，另一种则按文章脉络进行概述。

2. 写作用语

结论的撰写有如法律条文，表述的形式和使用的字眼必须明确、精炼、完整、准确，不得含糊其词或模棱两可，文字上也不应夸大，对尚不能完全肯定的内容注意留有余地。例如，对不能明确的或无确切把握的结论，可用"印象"二字表示，并适当选用"看来""似乎""提示"等留有余地的词，以代替"证明""证实"等肯定的词。此外，应避免使用口头语言和过于感情化的用语。

3. 内容安排

结论中可以包括必要的数据，但不宜过多，应该明确自己的见解和观点，不宜引证他人的理论或尚未明确的问题，也不必要叙述有关观点争论的细节。结论必须与材料、结果、讨论相呼应，后者是前者的支持和保证，结果中没有的内容或是讨论中未提及的问题，都不得在结论中出现。应该避免在结论中简单地重复前文中的描述或是研究成果，对于与本研究无关或是不重要的内容，应该删去。

5.5.3 语态和时态的运用

结论部分的语态和时态表达与讨论部分基本相同,若描述所做的研究工作通常用一般过去时,描述普遍现象通常用一般现在时,有时也用一般过去时。结合具体内容,可分为:

(1)过去时。①描述本研究的内容;②描述他人研究的内容;③作者认为只适用于本研究环境和条件的结论。

(2)现在时。①指示性说明;②普遍接受的思想、理论或结论;③作者认为具有普遍意义的结论。

5.6 正文结构化分析

为充分表达研究成果的来龙去脉,提高论文的可读性,在写作过程中一定要注意合理安排论文的篇章结构,力求论文的结构严谨、内容充实、论述完整、逻辑性强,在研究背景的介绍、学术思想的解释等论述中,要尽量做到深入浅出,表达清楚、简洁,专业术语准确且前后一致。

近年来,有关科技论文的结构化表达日益受到期刊编辑和研究人员的重视。例如,鉴于结构式摘要的高效和规范,英国 Keele 大学心理学教授 Hartely 博士于 1999 年提出的"结构式论文"(structured article)的概念,并详细阐述了结构式论文中各级标题的名称、构成及内容[13,14]。又如,Docherty 等[15]认为,讨论部分是论文的卖点,应该强调该部分的规范表达,并建议科技论文中讨论部分要结构化,讨论的结构和内容应涵盖以下诸方面:重要结果(或发现)的概述;本研究的进步与不足;本研究成果的特色(与他人成果比较而言);研究意义;存在问题及进一步研究的方向。

目前在论文结构化方面做得比较成功的是"报告试验的统一标准"(consolidated standards of reporting trials,CONSORT)。CONSORT 自问世以来得到了越来越多的期刊和国际编辑学组织的支持,并分别于 1999 年和 2000 年修订[16-18]。2000 年修订 CONSORT 对随机试验报告的篇章结构及各部分的具体内容和描述均有较明确的规定,其中的"随机试验报告项目核查表"包括了五个部分(共计 22 项),内容大致为[17,18]:

(1)题名和摘要。参与者如何被分配到干预组[如随机分配(random allocation,randomly assigned)、随机(randomized)等]。

(2)引言。研究背景和理论基础。

(3)方法。参与者、干预措施、目的、测量结果与测量方法、样本量、随机化、盲法、统计学方法。

(4)结果。参与者流程、募集、基线资料、受试者的数量分析、结果与估计、辅助分析、不良事件和副反应。

(5)讨论。解释、可推广性、根据现有的证据解释结果。

实际上,有些知名期刊,如 *Nature*、*BMJ* 对其所刊载各栏目论文的结构安排、甚至论文中各部分的篇幅(或大约数字)均有较为严格的规定。Nature 的"作者指南"特别指出,如果作者不遵循有关论文的体例结构与内容方面的规定,有可能会导致编辑部对稿件处理的严重延误。

本章参考文献

[1] 佚名. 科技论文写作整理系列(5):materials & method[EB/OL]. [2021-09-10] http://www.dxy.cn/bbs/post/view?bid=45&id=4540956&sty=1&tpg=1&age=0

[2] 任胜利. 英语科技论文撰写与投稿[M]. 北京:科学出版社,2004.

[3] KO J, JEONG S. Plugging effect of open-ended piles in sandysoil[J]. Canadian Geotechnical Journal,2015,52(5):535-547.

[4] GAUTAM T P, SHAKOOR A. Comparing the slaking of clay-bearing rocks under laboratory conditions to slaking under natural climatic conditions[J]. Rock Mechanics & Rock Engineering,2016,49(1):19-31.

[5] LAI H J, ZHENG J J, ZHANG J, et al. DEM analysis of "soil"-arching within geogrid-reinforced and unreinforced pile-supported embankments[J]. Computers and Geotechnics,2014,61:13-23.

[6] WU W B, El NAGGAR M H, ABDLRAHEM M, et al. A new interaction model for the vertical dynamic response of pipe piles considering soil plug effect[J/OL]. Canadian Geotechnical Journal,54(7),(2017-02-18)[2019-10-10]. http://doi.org/11.1139/cgj-2016-0309.

[7] JEONG S, KO J, WON J, et al. Bearing capacity analysis of open-ended piles considering the degree of soil plugging[J]. Soils and Foundations,2015,55(5):1001-1014.

[8] 佚名. 如何撰写英语科技论文之十:讨论(Discussion)[EB/OC]. [2020-10-19]. http://blog.sina.com.cn/s/blog_605a554d0100mv3i.html.

[9] HARTLEY J. To attract or to inform:what are titles for?[J]. Science Foundation in China,2004,12(z1):59-64.

[10] HARTLEY J. Abstracts, introductions and discussions:how far do they differ in style? [J]. Scientometrics,2003,57(3):389-398.

[11] ZHAN J W, XU P H, CHEN J P, et al. Comprehensive characterization and clustering of orientation data:a case study from the Songta dam site,China[J]. Engineering Geology,2017,225:3-18.

[12] LAI H J, ZHENG J J, ZHANG R J, et al. Classification and characteristics of soil arching structures in pile-supported embankments[J]. Computers & Geotechnics,2018,98:153-171.

[13] HARTLEY J. From structured abstracts to structured articles:a modest proposal[J]. Joural of Technical Writing and Communication,1999,29(3):255-270.

[14] HARTLEY J, PENNEBAKER J W, FOX C. Abstracts, introductions and discussions: how far do they differ in style[J]. Scientometrics, 2003, 57(3): 389-398.

[15] DOCHERTY M, SMITH R. The case for structuring the discussion of scientific papers[J]. British Medical Journal, 1999, 318(7193): 1224-1225.

[16] 汪谋岳. 如何报告随机对照试验: 报告试验的强化标准简介[J]. 中国科技期刊研究, 2003, 14(2): 222-224.

[17] MOHER D, SCHULZ K F, ALTMAN D. The CONSORT statement: revised recommendtions for improving the quality of reports of parallel-group randomized trials[J]. The Lancet, 357(9263): 1191-1194.

[18] 刘雪梅, 刘建平. 应用CONSORT提高医学期刊质量[J]. 编辑学报, 2002, 14(3): 228-229.

第6章　致谢与参考文献

6.1　致谢的必要性与内容

致谢部分作为学术论文主体内容外的单独部分，一般是作者用来对在研究进行和论文撰写过程中所获得的帮助进行礼貌性的感谢。致谢部分一般位于文章的结论之后，参考文献之前，其内容可参考以下几点[1,2]。

（1）Financial support(recognition of extramural or internal funding)。

（2）Technical or instrumental help (providing access to data, tools, technologies, facilities, and also furnishing technical expertise, e. g. , statistical analysis)。

（3）Conceptual (source of inspiration, idea generation, critical insight, intellectual guidance, assistance of referees, etc.)。

（4）Editorial (providing advice on manuscript preparation, submission, bibliographic assistance, etc.)。

（5）Moral(recognizing the support of family, friends, etc.)。

6.2　致谢的写作要点

6.2.1　致谢的内容应尽量具体和恰如其分

致谢中应列出对论文工作有重要帮助或特殊贡献的人或机构，应尽可能具体说明支持的事项或类型(技术支持、提供实验材料、稿件的修改、经济支持等)，例如：

The work reported in this paper was funded by China National Funds for Distinguished Young Scientists with Grant No. 51025932 and National Natural Science Foundation of China with Grant No. 51179128. The support of the Itasca Consulting Group is also gratefully acknowledged. Thanks also go to Dr. Banglu Xi from Tongji University, China, for his contribution on the paper. All the supports are greatly appreciated[3].

同时，为表示应有的礼貌和尊重，尤其当在致谢中提到人名时，需要提前获得此人的许可和其对致谢部分措辞的认可。

6.2.2 用语要恰当

致谢中用语的选择要恰当,避免因疏忽或用语不当而冒犯需要感谢的个人或机构,例如,在表达希望、愿望时,wish 很合适,但如果说"We wish to thank David",则是在浪费单词,且蕴含着"I wish that I could thank David for his help but it was not all that great."(我希望感谢 David 的帮助,但实际上他的帮助并不是特别大)的意思[4]。恰当的表达为"We'd like to thank David."。上述表达以及一些类似表达在致谢中很常见,这一点要尤其注意。

6.2.3 遵从期刊要求和相关规定

投稿前,需仔细阅读拟投期刊的"作者须知"(Guidelines)和该期刊上已发表论文的致谢部分,应遵循拟投期刊致谢部分的格式要求,例如基金资助的信息,有些期刊要求将这一部分放到论文首页的脚注中,有的则是放到致谢中,且不同期刊对资助项目号、合同书编号等基金信息的要求也不一样。

6.2.4 利益冲突声明

有些文章致谢部分的最后是关于利益冲突的声明。当作者(或作者所在机构)、评审者或编辑之间有影响他们行为的(偏见)经济或个人关系,就存在利益冲突。这些关系的程度各不相同,小到可以忽略它对判断的潜在影响,大到对判断产生极大的潜在影响。同行评审和出版过程中的所有参与者都必须公开所有可视为潜在利益冲突的关系。这种关系的公开对编者按和评述文章来说尤其重要,因为在这类文章中发现偏见比在原创性研究论文中更困难。

关于利益冲突声明的写法可采用以下几种方式[4]:

(1) The authors declare no competing financial interests。

(2) No potential conflict or interest relevant to this article was reported。

(3) A university owns a patent, xx, yyy, zzz, that uses the approach outlined in this article and which has been licensed to B。

(4) The authors have nothing to disclose。

6.3 参考文献的重要性

科技论文正文后面所附参考文献指的是撰写论文时引用的有关文献资料。在写论文时,凡是引用了他人、前人或作者本人已公开发表的观点、数据和资料等都需要在论文中加以标明,并在文后列出,这一过程称为参考文献著录。参考文献是一篇完整的科技论文中不可或缺的一部分,往往是作为论文正文部分的某种缘起及延伸的介绍,同时也可引领读者获取其感兴趣文献的信息。另外,引用他人文献是对他人贡献的一种承认,参考文献还给编辑提供了审稿人信息,并显示了作者对本专业领域的熟悉程度。因此,参考文献部分的内容同样具有重要意义。

一般来说,参考文献的作用主要有:①说明研究工作是在怎样的背景下产生的(即研究背景、依据和历史渊源);②佐证与前人研究工作的异同,避免作者重复介绍已有的研究方法、结果和结论;③综述或专著材料的来源(应遵照版权法合理使用);④提供给编辑和读者查阅引文的途径;⑤编制引文索引和引文统计(计算影响因子)的数据来源和基础。

6.4 参考文献的选取原则

对参考文献的选取能很好地反映作者对所研究领域的熟悉程度,甚至在一定程度上能反映作者的学术水平和科研态度。因此,引用时应尽量选择最重要的、水平最高的和与研究主题联系最紧密的论文,而不是列出与某个话题相关的所有文献,这样会加大读者的阅读障碍,同时让审稿人产生作者对科技论文写作不专业的印象。对参考文献的选取可遵循以下原则:

(1) 所引用参考文献的主题必须与研究内容密切相关,当相关领域研究成果较多时,可适量引用高水平的综述性论文以对相关领域进行概括;同时,当文献的参考价值相同时,优先选择最新发表的成果。

(2) 所引用的文献必须是亲自阅读过的,若引用的是二次文献(从某篇论文的引文中所获得的文献),且原文献无法得到,应在正文中指出是从那篇文献转引过来的。

(3) 应尽可能引用已公开发表且便于查阅的文献。当文献的参考价值相同时,优先选取高水平期刊上发表的论文;对于还未公开出版或其他形式的参考文献,引用格式应遵循拟投稿期刊的相关规定;对于期刊不建议的文献形式,若文献资料对论文内容有较大作用,仍可在论文正文中提及。

(4) 应尽量将文献数量控制在合适的或期刊要求的规模,同时避免不必要地、过多地、目的性地引用作者本人的研究成果(尤其当该研究成果与现研究主题联系不紧密时)。

6.5 文献在正文中的使用

科技论文正文中需要引用参考文献的情况:

(1) 引言。只引用最相关的文献。引用最近的、最重要的、最相关的、尽可能接近第一手的文献。当相关领域研究成果较多时,可考虑引用水平较高的综述性文章。切忌堆积参考文献,这样会导致读者难以发现关键信息。

(2) 材料与方法。为研究中使用的材料或方法引用第一手文献,包括发表在高影响力期刊上的方法,而不是详述那些方法的细节。作者自身所做的工作才是需要重点介绍的。

(3) 结果。需要引用文献的陈述通常不在结果部分,而在讨论部分,例如与前人研究成果的对比。但是,如果一个简单的对比不适合在讨论中出现,也可以写在结果中,这样就需要引用文献。注意只引用文献中的关键信息,主要的关注点还是作者本身的研究成果。

(4)讨论。尽管研究结果是主要话题,但需要在一个广泛的范围内来讨论结果。这意味着需要引用文献来对比研究结果,参考其他研究对结果的解释,或借用其他文章来说明结果的重要性。

在科技论文正文中引用参考文献时,需要注意下面几部分的内容。

6.5.1　切忌抄袭原文

在论文写作过程中,对他人成果加以引用而不注明信息来源的做法会被认为是抄袭。学术抄袭是一种学术不端行为,涉及学术伦理、学术规范,甚至知识产权,因此,对这一行为应绝对避免。在进行科技论文写作时,避免不了对他人工作的引用和改写,但需要注意改写与抄袭的差别。改写是运用自己的语言对他人的思想进行表达与借鉴,而只替换他人成果中的一两个词或是改变句子结构不是改写,是抄袭。下面给出一段改写与抄袭的例子[4]。

原文:It has currently been recognized that both the type and characteristics of the rust layers formed on the steel surfaces are very important because they can determine their protective properties. According to a recent theoretical model developed by Hoerlé et al.[2], the long-term corrosion behavior of iron exposed to wet-dry cycles is largely controlled by the characteristics of the rust layers. Additionally, the differences between the corrosion behavior observed for different types of steels have been related to the rust layer characteristics. Okada et al.[8] have carefully investigated, by using detailed variable temperature Mössbauer spectrometry, the protective rust formed on both weathering and mild steels after 35 years of exposure to a Japanese semirural type atmosphere. They reported that the rust on both steels it composed of goethite(major component), lepidocrocite(minor component) and traces of magnetite. 。

构成抄袭的段落:Both the type and characteristics of the layers formed on the steel surfaces are very important because they determine their protective properties. Recently, Hoerlé et al. developed a theoretical model[2] that the long-term corrosion by the characteristics of the rust layers. The differences between the corrosion behavior observed for different types of steels have also been related to the rust layer characteristics. Using detailed variable temperature Mössbauer spectrometry, the protective rust formed on both weathering and mild steels after 35 years of exposure to a Japanese semirural type atmosphere have been determined by Okada et al.[8] who reported that the rust on both steels is composed of goethite(major component), lepidocrocite(minor component) and traces of magnetite.

改写后的段落:Rust constituents can determine the performance of steels and influence their life expectancy. The characteristics of the rust layers control the long term corrosion behavior of iron exposed to wet-dry cycles[2]. For example, after 35 years of exposure in Japan, rust formed on steels under various conditions is composed of mainly goethite, some lepidocrocite, and traces of magnetite[8].

对比上述三段文字可以看出,构成抄袭的段落中,虽然作者已注明了参考文献,但作者除了改动几个单词,基本照搬了他人的工作,因此,这仍构成抄袭。在改写后的段落中,作者用新的语言对他人的工作进行了介绍,在结构和用语等方面都与第二段文字也形成了明显的差别。

在科技论文写作过程中,部分学者因对文献引用不够规范,会带来抄袭之嫌。为避免上述无心之失,可以考虑以下几点:

(1)对需要引用或改写的部分做一个概括,并详细记录文献的来源。
(2)若需要引用原文,需要让原文突出主体文字,例如改变字体,加引号等。
(3)对文章进行改写时,结合之前做的概括,用自己的语言对他人的工作进行叙述,并核对原文献以确保表述无误。

6.5.2 遵循拟投期刊应用格式

由于研究领域不同、所述体系不同等原因,不同期刊会存在不同的文献引用格式,大部分期刊在其主页上都会有"Guide to Authors"的部分,对作者的投稿提供指导,包括参考文献的引用格式。这里主要介绍两种较为常见的文献引用体系。

1. 作者-出版年体系(name-year system)

作者-出版年体系的要求是,正文文献的著录由作者的姓氏和出版年份构成,而参考文献列表中则按作者姓氏的首字母顺序和出版年份的先后来确定。采用作者-出版年格式时,由于不需要对文献进行编序,因此无论增加或删减文献,文献的引用格式都不需要改变。

这种格式同样存在一些不足。当文献引用量较大时,读者往往需要花费一定时间确定是具体的文献。同时,大量文献的插入也会影响读者阅读的流畅性。

使用作者-出版年体系时需注意以下几点:

(1)若引文中没有直接体现作者姓氏,则需要单独加括号引用;若在正文中体现了作者姓氏,则需在括号中提供出版年,例如,"Shanley and Mahtab(1976)proposed a counting method using stereographic projections to identify the joint sets."[5]。

(2)若引用同一作者在同一年份发表的多篇论文,需要在括号中的出版年后加字母(a、b、c 等)加以区别,例如,"Kulatilake et al.(1995a)suggested that…"。

(3)当参考文献的作者小于或等于 2 位时,则所有作者的姓氏均应在正文中列出;当参考文献的作者大于或等于 3 位时,正文著录只需列出第一作者加 et al.(具体作者数目依据拟投期刊的要求),例如,"Examples include rock mass characterization(Kulatilake et al., 1997),earthquakes(Malamud and Turcotte,1999),drainage networks(Fac-Beneda,2013),soil science(Montero,2005),and geological disasters(Sezer,2010)."[5]。

2. 顺序编码体系（citation-order system）

顺序编码体系的要求是，文献按其在正文中出现的先后进行排序，并将序号置于方括号"[ ]"或圆括号"()"中，而在文献列表中将各参考文献按它在正文中出现的序号依次编排。当引用多篇文献时，只需将对应文献的序号在括号内列出，各序号间用","隔开，对连续序号可采用起止序号，例如，"The third group assumes…；there is no consensus among practitioners on the appropriate soil arching model to use for practical design."[6]。

与作者-出版年体系相比，顺序编码体系更加简洁，且方便读者更快地在文献列表中找到对应的文献；但在进行文献的增加或删减时，可能会导致参考文献序号的混乱。

6.5.3 用语恰当、礼貌

引用文献的目的有支持、反驳、对比、突出作者的思想或研究成果等目的，无论出于何种目的，引用他人的研究成果时时务必要专业和谦恭，尤其是评价他人研究的不足或欠缺时，不要用侮辱或绝对否定的词汇。不礼貌的表达可能会令编辑或审稿人不愉快，从而导致文章被拒。例如，"The study by James is without merit."，应修改为"Our study differs from that of James in that…"。

下面给出一些引文中常用的礼貌性表达：

(1) 表示支持相关研究成果，"Similar results have also been observed by Alton et al. (2005)."。

(2) 反驳他人的研究成果，"However, evidence such as that provided by Lim and Lehane(2015b) and Carroll et al.（2017）, indicating that little or no set-up occurred in certain circumstance."[7]。

(3) 与他人成果进行比较，"This trend is consistent with the mechanisms of friction fatigue presented by White and Lehane(2004), and others."[7]。

(4) 突出作者本人的成果，"Unlike other previous findings, our work presents…"。

6.6 文献在列表中的编排

进行文献引用时，除了要在正文中提供参考文献的主要信息，还需在"References"部分提供所有文献的文献列表，列表中应包含参考文献的各项基本信息，如作者姓名、论文题目、期刊的名称、年、卷、期或专著的出版年、出版地、出版社、起止页码等。一些期刊还要求在参考文献后提供相关链接，例如 DOI 号，以方便文献的核对与搜索。为了满足不同期刊对参考文献的要求，近年来开发了各类型的文献管理软件，如 Mendeley、Zotero、Endnote 等，均能满足基本的文献编排需要。在进行具体的参考文献编排时，一定要仔细参考拟投稿期刊的格式要求。下面对文献在文献列表中编排的注意事项进行介绍。

6.6.1 作者姓名

作者姓(surname 或 family name)和名(first names)的正确识别,对于文献列表中参考文献的排序、文献检索等非常重要。文献列表和论著索引中通常都采用姓前名后的形式(姓氏的字母全部拼写,名缩写为首字母),不同期刊中对文献目录中作者姓氏识别的要求在细节上可能稍有不同,需参考拟投期刊的要求。

6.6.2 文献题目的格式

文献列表对于不同类型参考文献题目著录格式的要求不一定相同,作者应根据具体期刊的要求来撰写。一般来说,对于科技论文,论文题目需要首字母大写,其他单词一律小写(专有名词除外),例如,"Fac-Beneda,J.,2013. Fractal structure of the Kashubian hydrographic system. J. Hydrol. 488,48-54."。至于专著类文献,要求各实词的首字母均采用大写形式,例如,"Mandelbrot,B. B.,1982. The Fractal Geometry of Nature. W. H. Freeman,New York."。

6.6.3 期刊名缩写

为了节省文章篇幅,目前许多期刊都会要求文献列表中的期刊名称采用缩写形式,期刊的投稿指南中一般会给作者提供"journal abbreviations source"进行参考。若对部分期刊名的缩写不确定且无法查证,建议在投稿时对期刊名统一采用全称,并在文章被录用后可按期刊编辑的建议进行修改。目前,对期刊名进行缩写已成为一种趋势,因此,了解一些基本的期刊名缩写规则对参考文献的引用和检索都是十分有帮助的,例如:"Journal"只能缩写为"J.",所有的"-ology"都缩写到"l"等,比如"*Engineering Geology*"缩写为"*Eng. Geol*"。

6.6.4 期刊或专著的出版时间

期刊文献的出版年是必不可少的,有些期刊还要求加注卷号(Vol.)和期号(No.),期号通常置于圆括号内,例如,"Ono K,Yamada M. Analysis of the arching action in granular mass. Géotechnique,1993,43(1):105-20."。对于文章已被期刊接收但未确定出版时间,引用时需要标注"In Press"。

对于专著来说,除标注出版年外,还应标注出版版次,第 1 版不著录,其他版本说明应著录。版次序号通常采用序数缩写的形式(如 1st ed.、2nd ed.、3rd ed.、4th ed. 等)。

6.6.5 论文的起讫页码

论文的起讫页码能帮助读者直观地了解到该文献的内容量,是 1 页的 note 还是 30 页的综述,从而间接地判断文献中可能包括的信息量,方便读者进一步地确定是否需要继续对该文献进行追踪和获取。下面提供一些通用的页码格式规定:

(1)对于99以内的第2个数字(终止页码),第2个数字要写全,如2-5、9-13、89-99等。

(2)对于较大的数字,第2个数字(终止页码)只需写出最后2位,但有必要时应多写,如96-101、923-1003、103-04、1003-05、395-401、1608-774等。

6.6.6 常用的文献编排格式举例

本节结合爱思唯尔(Elsevier)多媒体出版集团旗下期刊 *Computers and Geotechnics* 所提供的"AUTHOR INFORMATION PACK"[8]及该期刊已发表论文的文献格式,对一些常见的文献编排格式进行举例。

1. 期刊论文

(1)一般引用,如"Van der Geer J, Hanraads JAJ, Lupton RA. The art of writing a scientific article. J Sci Commun 2010;163:51-9. https://doi.org/10.1016/j.Sc.2010.00372."。

(2)文章号引用,如"Van der Geer J, Hanraads JAJ, Lupton RA. The art of writing a scientific article. Heliyon. 2018;19:e00205."。

(3)非英文论文,"Xia YY, Rui R. Experimental analysis of vertical soil arching effect of embankment reinforced by rigid piles. Chin J Geotech Eng 2006;28(3):327-31. (in Chinese)."。

2. 专著及相关

(1)专著。如"Terzaghi K. Theoretical soil mechanics. New York:Wiley;1943."。

(2)会议论文。如"Guido V A, Kneuppel J D, Sweeney M A. Plate loading test on geogrid reinforced earth slabs. In: Proceedings of Geosynthetics'87, New Orleans, USA. IFAI;1987. p. 215-25."。

(3)学位论文。如"Demerdash M A. An experimental study of piled embankments incorporating geo-synthetic basal reinforcement. Doctoral Dissertation, University of Newcastle-Upon Tyne, Department of Civil Engineering;1996."。

(4)行业规范。如"German Standard, EBGEO. Recommendations for design and analysis of earth structures using geo-synthetic reinforcements. ISBN 978-3-433-60093-1. German Geotechnical Society;2011."。

(5)合同书。如"Filz G M, Smith M E. Design of Bridging Layers in Geosynthetic Reinforced Column-supported Embankments. Contract Report VTRC 06-CR12. Virginia Transportation Research Council, Charlottesville;2006."。

3. 其他

(1)软件。如"Itasca Consulting Group, Inc. PFC2D user's manual, version 4.0. Itasca

Consulting,Group,2008."。

(2)网络材料。如"Cancer Research UK. Cancer statistics reports for the UK,http://www.cancerresearchuk.org/aboutcancer/statistics/cancerstatsreport/;2003［accessed 13 March 2003］."。

(3)数据库,"Oguro M,Imahiro S,Saito S,Nakashizuka T. Mortality data for Japanese oak wilt disease and surrounding forest compositions,Mendeley Data,v1;2015. https://doi.org/10.17632/xwj98nb39r.1."。

本章参考文献

［1］HARTLEY J. From structured abstracts to structured articles:a modest proposal［J］. Journal of Technical Writing & Communication,1999,29(3):255-270.

［2］佚名. 如何撰写英语科技论文之十二:致谢(Acknowledgement)［EB/OL］.［2020-10-19］. http://blog.sina.com.cn/s/blog_605a554d0100mv3m.html.

［3］JIANG M J,DAI Y S,CUI L,et al. Investigating mechanism of inclined CPT in granular ground using DEM［J］. Granular Matter,2014,16(5):785-796.

［4］任胜利. 英语科技论文撰写与投稿［M］. 北京:科学出版社,2004.

［5］ZHAN J W,XU P H,CHEN J P,et al. Comprehensive characterization and clustering of orientation data:a case study from the Songta dam site,China［J］. Engineering Geology,2017:225.

［6］LAI H J,ZHENG J J,ZHANG R J,et al. Classification and characteristics of soil arching structures in pile-supported embankments［J］. Computers & Geotechnics,2018,98:153-171.

［7］ANUSIC M I,LEHANE B M,EIKSUND G R,et al. Evaluation of installation effects on the set-up of field displacement［J］. Canadian Geotechnical Journal,2018,56(4)(2018-06-13)［2020-05-06］. http://doi.org/10.1139/cgj-2017-0730.

［8］Elsevier. Author information pack［EB/OL］.(2018-09-29)［2020-05-20］. http://www.elsevier.com/locate/compgeo.

第7章　论文图表

7.1　概　述

图表是现代科技文献的重要组成部分[1,2]，是表达实验数据直观而简洁的方式。统计显示，在现代科技书刊中，平均每100字就接有一幅插图或表格[3]。要从浩瀚的文献中进行选择性阅读，提高阅读效率，读者常常在完整地阅读一篇科技论文之前，除了先阅读题名和摘要之外，更会浏览文中的图表来获取主要数据和把握论文的主要思想，从而决定是否精读这篇论文。

一篇优秀论文的表达，除文字叙述外，图和表的运用不容忽视。只有文字、图、表三者的有机结合和互相补充，才能使论文达到结构严谨和逻辑性与可读性强的目的[4]。

7.2　图表的作用与分类

7.2.1　图表的作用

科技论文的目的是描述实验(试验)和研究的技术方法、论证过程、分析结果和所得结论，往往涉及观测和实验数据、定量化信息的计算分析与比较等。表格能够准确地记录和提供关键的数据、定量化论据和结果，而且能够准确地表达内容的比对和逻辑关系。图形和图像则可以形象、直观地表达文字难以描述的科学思想和技术知识。此外，规范和正确地使用图表不仅可以大量省略文字、紧缩篇幅、节约版面，而且可以活跃和美化版面，使读者赏心悦目，提高阅读兴趣和效率。可以想象，一组定量化论据，如果采取数据表格的形式插入科技论文中，则可以非常准确、明了地呈现给读者，而且数据间相互紧凑，便于比较其大小、分布和相互关系；地图、大气环流形势、气象要素分布、机械构造、电路、生物形态和结构、工作或信息流程等，即使采用大量篇幅的文字描述，读者恐怕也难以准确、清晰地了解它们，而采用图形或图片(照片)则非常直观而简洁。因此，图和表是文字的两个"翅膀"，是现代科技论文和科技书刊不可缺少的表述手段[3]。

总的来说，图、表在科技论文中的作用可归纳为以下几点[5]。

(1)证明作者的工作进展。论文的目的是力求说服读者了解作者的研究工作，承认作者研究工作的真实性。假如没有图表,读者不清楚作者是否做了实验,做了哪些实验,如何做的实验,难免对作者写的论文持有怀疑态度。例如某一作者说"该实验设备操作简便、携带方便、实验结果稳定可靠,有极大的推广价值",读者可能会对这文字描述半信半疑。如果作者描述"该实验设备照片如图 1 所示,质量仅为 500g,可用于现场便携使用,实验结果如图 2 所示",读者一看到图片就会相信作者的论点,知道作者的实验设备距离推广应用已经不远了。

(2)提供第一手数据。图表是忠实记录测试结果的原始数据,这是科学研究的第一步。有了图表所提供的实验对象、条件和结果,作者才可能展开进一步的议论。例如,作者写道"采用这种加速度传感器采集的信号稳定可靠,可满足实验需求",却不提供任何测试结果,就显得有些空口说白话,无法令读者相信。如果作者写的是"为了验证该加速度传感器的采集效率,先后对尼龙锤、铁锤激振产生的加速度信号进行了采集,如图 1 和图 2 所示",有相关采集图形作为支撑,读者一下子就明白了作者论述的合理性。值得注意的是,绝不能形而上学地预期实验结果,去凑数据,科学研究的魅力就在于有些科学数据中隐含的真理可能在今后数十年甚至数百年才能被人揭晓。

(3)表明作者的科研能力。图表应符合所投期刊的要求,且应追求精致美观。图表制作能反映一个作者的素养和水平。和读者一样,审稿专家也喜欢看精致、美观的图表。好的图表可以给审稿专家留下很高的印象分,并吸引专家以欣赏的态度审稿;相反,较差的图表会大大降低论文的水平和层次。

7.2.2　图表的分类

科技论文中的插图包括曲线图、记录图、结构图、流程图、框图、布置图、地图、照片、图版等[5,6]。其中,常见的曲线图有函数关系曲线图、等值线分布图,例如,大气科学相关领域研究中常使用的气象要素时间变化曲线、空间分布曲线,以及地面气压场、位势高度场、气象要素或物理量场时间或空间剖面图等;结构图多用于呈现某一物体、机械部件的全部或部分详细构成;示意图不详细刻画物体细节或数量的微小变化,而只需呈现它们的总体形势和变化分布形态[7];流程图和框图形式相似,它们多半都由多个文字框、符号框或数据框组合构成,不同的是流程图侧重于表达事物演变和变化过程、工作或工程的步骤和顺序、信息传递方向等,而框图侧重于表达事物的构成、布局;照片则主要用于反映物体,特别是生物的外貌形态和特征。

表格按照其内容和编排形式主要分为数据表和系统表[8]。其中,数据表是科技论文和科技书刊中最常用的。它是以卡线表形式列出有关数据或数量,有些非数据型的符号、公式、表达式或论文中需要说明的事项也可以采用卡线表形式列出;系统表一般用于表示多个事项之间的隶属关系和层次,有时也采用类似汉语主题词表的编排格式把事项的层次关系、属性等列出。

7.3 插图的设计制作

7.3.1 插图的设计制作原则

科技论文的插图应当具有自明性、写实性、规范性和示意性。

(1)自明性。读者只看图题、图面和图例即可完整无误地理解图意,而无需阅读正文。

(2)写实性。严格地忠实于描述对象的本来面貌,只要是对表述论文思想和对读者有意义的信息都要清晰完整地呈现出来,没有模糊的描绘,更不可以臆造、添删或改动。美术作品可以运用虚实结合、夸张等人工处理手法,使部分画面模糊虚脱,以重点刻画主体或物象的主要部分,而科技论文的插图是决不允许的。

(3)规范性。规范性有两重含义:①插图应当重复地、规格化地设计和描绘,使科技信息准确、方便、直观且唯一地反映给读者。一切含义晦涩、需要读者深思才能逐步发掘的设计和制作方法都是不可取的,更不允许存在多重含义或需要读者凭想象理解可能产生歧义的内容。这可能是科技论文插图不同于工艺图画的主要特征。通常工艺图画都强调求新出奇,重复的规格化的设计和描绘是低廉的工匠手法。②插图应当构成要素完整,图面布局和结构符合国家标准或专业技术标准及科技书刊标准化的编排要求[9-11]。

(4)示意性。科技论文的插图是与文字配合的,是辅助文字描绘文字难以表述的内容,所以,插图应当简约,突出主题。①科技论文中的插图不等同于工程、工业制图和供设计或计算用的图纸、技术手册,一般不需要标注尺寸比例(某些有尺寸比例效应的图例外),部分附注技术条件等文字说明也可以省略。②插图应当精选。论文中只使用具有典型意义、反映主题内容所必须的插图。

当然,科技论文的插图设计和制作也应当清晰、美观,画面、线条在能够准确表达主题内容的前提下力求简洁、明快。

7.3.2 插图的制作

1. 插图的构成要素

科技论文的插图由图号、图题(名)、图面三个部分构成[3]。图号采用阿拉伯数字按插图在文献中出现的先后顺序连续编号。图题采用简明确切的文字和符号表述图面的主题内容及其属性。图号和图题置于图面下方。必要时,可以将多个图面连续摆放,使用同一图号和图题构成组合图,以共同反映同一主题。此时,必须在每幅图面下方标注分图序号,如(a)、(b)、(c)等,而且应当在分图下方或在图题中标注分图图题。

图题是插图的名字或标题,等同于论文的题名,当然应遵循题名的写作要点[12],用准确、恰当、简洁的词汇和短语说明图面的主题内容和属性,但不可过于笼统、模糊和泛指,如"函数曲线图""框图""示意图"等。

2. 函数关系图的制作

(1) 坐标的标目和标值。函数关系曲线图是地质工程及其相关领域科技论文中最常用的插图,图面由两个或两个以上坐标轴和函数关系曲线构成。坐标轴由轴线、增值方向、标目、标值和标值短线构成。标目包括坐标名称、量符号(坐标代表的量名称和符号)和单位,必须标注完备,以保证插图的自明性。坐标名称和单位之间用"/"分隔,标目与所标注坐标轴平行同向居中书写,位于坐标轴和标值外侧,纵坐标的标目遵循"左顶底右"的原则书写,顺时针旋转 90°为字符正向。标值应疏密得当,标注在坐标轴外侧并紧靠表值短线,表值的数字应尽量不超过 3 位,同一坐标轴的表值应具有相同精度(小数位数相同),小数点后不超过 1 个"0"。因此,应当按标准使用标准词头改造数量的量级或进行单位换算[13-15],如用"30km"代替"30 000m",涡度为"0.000 056s^{-1}"应当标注为"5.6",并在标目中标为"$10^{-5}s^{-1}$"。

(2) 画面的覆盖率。函数关系曲线图的坐标轴长度和标值范围应适当,既要包含必要的曲线,又不能太长或标值范围过大,使图面留有大片空白;要恰当设计坐标轴的相互位置,坐标轴的起点(交叉点)可以不是"0",避免图中空白过多[16]。各坐标轴长度的比例应得当,使函数曲线起伏明显,达到既可反映物理量之间的相互关系又协调美观的效果。

(3) 坐标的充分利用和图形的叠置。为了增强对比效果和节省版面,最好把一个参变量与数个物理量的多条函数曲线叠置在同一幅图上,每条曲线使用反差明显的不同线形描绘,并在图题或用图例标明各曲线代表的量名称或量符号和单位;当各物理量的量级差别较大,或曲线过多挤在一起难以分辨时,可使各物理量函数曲线分立于各自的坐标,即将几条曲线分散叠置在一幅图上。

3. 等值线图的制作

等值线图是灾害调查与防治等领域的科技论文最常用的插图之一。等值线图描述一个或几个物理量在一定时空范围内的分布状况,如地面气压场分析图,标准等压面高度场分析图,气温、风速、湿度、雨量、涡度、散度、垂直速度等分布图和垂直剖面图(空间或时间剖面)等。等值线图的图形范围应当能够反映所描述物理量的变化和分布特征,但为了突出重点和避免版面浪费,图形范围不可过大,例如,描述短期天气过程环流形势不应使用整个欧亚大陆甚至北半球高度场等值线图。等值线图中为了表明地理位置应在图边框上标注经、纬度,或者在图形下叠加地图(底图),经纬度表注在图形边框外侧且大小适当。地图一般采用只有海岸线、大的岛屿及主要河流和湖泊的简易地图,并使底图线条比等值线细而淡。等值线一般采用等间隔取值,特殊需要时可以采取变间隔取值描绘,取值间隔要适当,既保证线条疏密得当和清晰,又能够描述出物理量分布特征。线条应当平滑、粗细适宜,使图形美观。如果多个物理量等值线叠置在一张图上,则应采用不同线形(如实线、虚线、短线、点划线等)表示不同物理量,并在图题或图例中说明各线条的意义,不可仅采取不同颜色或灰度的线条来区分。等值线标值应当准确、清晰、大小适当、精度一致,有时为了清楚反映特殊量值范围,可以加灰色润饰,但不宜采用大面积深色,特别是应防止大面积涂黑。

4. 流程图和框图的制作

流程图和框图的设计应当布局合理、层次关系清楚、形式美观,如将如图 7-1(a)所示的布局修改成如图 7-1(b)所示的那样[3]，其传达的信息丝毫没有减少，不仅避免了图中的大块空白，还节约了将近一半的版面，而且外观更加整齐美观。此外，流程图和框图中应当使用本学科或专业领域熟知和共用的名词术语、量名称和符号，文字和符号等应准确、简洁，删去一切正文中没有说明或与正文内容不相关的文字、数字和符号，保持与正文一致。框图一定要层次分明、隶属关系明确；流程图必须用箭矢标明顺序和流向。

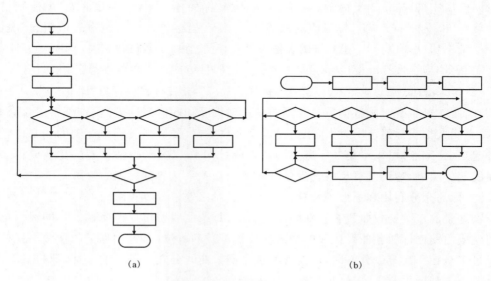

图 7-1　流程框图布局示例图
(a)布局一；(b)布局二

无论是哪种插图，为了图面清晰、简洁、美观，都应当去掉不必要的辅助线，不同函数关系曲线应使用不同线形的黑色线条描绘。

7.4　表格的设计制作

7.4.1　表格的设计制作原则

表格的设计和编排应当遵循科学、明确、简洁、自明和规范的原则。

(1)表格要有明确的表达目的性。把条件、方法、测试(实验或计算)数据、分析结果等分列清楚，使读者一目了然，而且排列应当使相同或相近属性的内容相互并列，便于比较和逻辑比对。

(2)表格应当简洁，只列出重要的现象、算式、参数、数据等，排除分析或计算的中间步骤或环节。

(3)突出重点。一般不列出调查、实验、分析计算使用的仪器、手段、材料或程序等事项说明。

7.4.2 表格的制作

(1) 无论数据表格还是系统表格都由表号、表名(表题)和表身组成。每一个表格应有简明、确切的表名,连同表号一起居中置于表身之上。数据表格的表身应尽可能采用三线表(无竖格线,左上角栏头中无划斜线),而不采用传统的卡线表。栏头中排列各项目名称作为栏目,栏目可以横向(表4-1)排列也可纵向排列(表4-2)[17],栏目应当使用国标推荐或本专业领域熟知或共用的名词或量名称、量符号,准确表明该栏的项目名称,量名称和符号后用斜杠"/"与单位连接,如气温 $t/℃$、气压 P/hPa。如果整个表格列出变量的单位相同,则栏头中的单位可以省略,把共用单位表注在表身右上端。采用国际标准计量单位和符号,单位标注应规范,如 $m·s^{-1}$、$g·kg^{-1}$、K 或 ℃、hPa、dB、dBz。

表 4-1 横表

××	×	×	×	×
×××	××××	××××	××××	××××
×××	××××	××××	××××	××××
×××	××××	××××	××××	××××

表 4-2 竖表

××	×××	×××	×××
×	××××	××××	××××
×	××××	××××	××××
×	××××	××××	××××

如果单纯的三线表不能清晰地表达要提供的信息或者容易产生数据混淆,则可以添加辅助线。表头上的辅助线一般用于区分栏目多层次的隶属关系,将同属一个栏目的多个分栏目放在一条短线之下;表身上的辅助线一般用于划分众多数据的不同隶属关系或数据簇,避免混淆和便于阅读查找。

(2) 表格中的数字应当准确、规范。同栏目中的数字应当精度一致;纯小数前的"0"不能省略;同类型数字需上下小数点对齐;上、下栏或左、右栏内容相同时要重复填写,不得使用"同上""同左"等字样;表内空白代表未测或无此项时,不可使用"—"或"…"代替,"0"代表实测或计算结果的确为 $0^{[18]}$。

(3) 当表格中某些信息需要注释时,可以在表格右端加备注栏或在表格底线下加表注,表格中需要注释的内容用阿拉伯数字加圆括号在该内容上加注上标,并与表注序号相对应。

(4) 对主要表述隶属关系的多层次事项,应采用系统表的形式。系统表格的布局应依人们的阅读习惯,自上而下或自左向右排列,编排要层次分明、隶属关系明确,布局要整齐、美观、节省版面。表中的文字、符号力求简洁、含义确切,避免使用大段文字或非共知的量符号组成的表达式。

7.5　图表的使用和常见问题

7.5.1　图表的使用

1. 恰当地选择插图与表格[3]

当用文字难以简洁、清晰、准确地表述科技论文中的内容时,需要使用插图或表格。某一内容是选用插图还是选用表格进行描述应根据插图和表格的特点,视描述的内容而定。插图的特点是形象而直观,表格则准确、逻辑对比性更强,所以,如果需要强调事物的外貌形态或变量的变化规律和趋势、分布形势等,宜选用插图,而如果需要准确地呈现测定数据、定量化论证和分析依据,以及要对比事物特征量、逻辑或隶属关系,则更宜选用表格。

2. 精选图表[3]

在科研论文的撰写过程中,为了观察、分析、剖析问题,往往需要收集或制作大量的图形、图像、照片和表格,但这些图表不能全部用到科技论文中。为了突出重点、节省版面,论文中的图和表必须精选,不使用可有可无或含义雷同的图表,只选用能够反映事物主要特征和对论点起关键支持作用的图表。对于反映事物演变的一系列图形、图像,则选取典型的具有代表性的一幅或几幅作为论文插图,也可以将这一系列图形的量值进行合成处理,制作成反映其普遍和平均特征的合成图。

表格中应尽量使用数据(观测、实验、计算结果)或等级,减少文字描述和说明,不仅增强表格内容的准确性、逻辑对比性,而且可以压缩表格空间。尽量合并同类型的表格。

要注意的是,避免使用插图和表格重复反映同一组数据和描述相同的主题内容。

3. 图表的放置和文字表述[3]

科技论文中的图表应当大小适宜。一般科技期刊版式为两栏,论文插图应尽可能采用一栏宽度,较大的插图可以采用两栏宽度,尽量避免使用折页图(幅面超过一页)。经适当缩放至大小适宜的图表应保证线条、标值、文字清晰,比例协调。

图表在论文中的放置位置遵循"图表随文"的原则,即先见文字表述后见图表,但图表应尽量靠近对它的文字表述,一般放置在文字段落后,这样既符合阅读顺序,又便于文字与图表的对照。

科技论文中的图表是辅助文字且准确、客观、真实、简洁地表述主题内容的,所以,所有的插图和表格在正文中必须有文字表述,说明图表内容的主要特点、属性,以及反映出的论证依据、规律性的结论,而且图表序号也必须在正文中至少出现一次。一是文字表述要与图表内容和特征相符,不可臆造或是图表中非典型、不显著的现象、特征;二是文字应尽量简练,不能成为图形图像原貌的文字描述,也不是表格数据的复述。

7.5.2　图表的常见问题

1. 图存在的问题[19]

科技论文中图的运用,使深奥的学术内容不再晦涩难懂,有助于读者对内容比较直观地了解。一个完整的图所包含的基本元素有图序、图题、标目、标值、图形、图注、图例等[20]。

(1)图序中的问题。编辑加工的稿件一般都是作者几经修改后的论文,可能存在删图的情况,但作者只在正文描述中做了相应的修改,图序却忘了改过来,导致缺图、少图及图文不一致的情况。

(2)图题中的问题。缺少分图图题是比较常见的现象,往往作者原稿里的一个大图里包含了几个小图,但都没有分图图题,有的虽然用(a)、(b)、(c)等序号加以区分了,但并没有相应的图题说明,缺少一定的自明性。另外,有些图题内容过于冗长复杂,缺乏简明性;一篇文章内同一试验不同试验条件所生成的图,图题却一样的情况也常可见;图题的中英文不对应,英文翻译存在语法错误和中式英语直译的现象。

(3)标目上的问题。常见的标目内容缺省不全和格式错误,如有的只画出了坐标轴和曲线,横坐标和纵坐标分别代表什么也没标注;有的虽然标注了,但只有量的名称却没有单位,或者标注的位置不对;量的名称字母缩写、大小写、正斜体、黑白体表述不规范,或者跟正文描述中的不一致;量与单位之间应用"/"隔开,可很多论文中仍用","或者"."隔开。

(4)标值的问题。标值有的疏密不当,过密的影响美观和简洁,过疏的缺乏对应的数据曲线,导致论据的说服力不够;标值数字过大,千分位未空格;标值的轴区间设定不恰当,取值不规整;已有标值的,函数坐标轴仍重复使用箭头,而流程图中却缺少箭头走向等。

(5)图形的问题。曲线对应的数值不对,比如说峰峰值、最高值、差值、相对值等在正文中的描述,在图中未能显现或者反映的趋势规律不一致;在有多条曲线重复的图中,因为分辨率不够,或者使用的线条区别不大(如不同粗细的线条),导致图片看上去层次模糊、线条混乱,起不到辨识对比的作用[21]。

(6)图注和图例的问题。并非所有的图都有图注和图例,如果图中的符号、缩写及图题中不能涵盖的必要信息,应加图注予以解释,所以不能是重复信息;而图例应该采用大多数文字处理软件都能编校和常见的标准符号,全文的图例应该统一一致,比如图1中图例A的曲线用圆点,图例B的曲线用方框,那么当图2同样表示A、B的曲线时,图例A采用短横线,图例B长横线,就是错误的。

2. 表存在的问题[19]

表格,简称表,是记录数据或事物分类等的一种有效方式,具有简洁、清晰、准确的特征,而且逻辑性和对比性又很强,因而在科技书刊和其他文献中应用广泛[22]。三线表因为保留了传统卡线表的全部功能,又比卡线表具有更强的直观性和简洁性,成为大多数科技期刊所采用的表格形式。

三线表主要包含表序、表题、项目栏、表身、表格线、表注。表序和表题的问题和图差不多,这里不作赘述。

(1) 项目栏。一般要放置多个栏目,要准确反映表身中该栏目的特征或属性。这一内容项必须精选,有些论文存在一些将不必要的项目列出来的情况,比如在使用表格说明燃料使用量同为 8L、不同材料的火灾实验数据时,燃油使用量这一栏就不必在表格项目中列出,在文中简单加一句说明即可;项目栏中在存在量和单位组成的数据时,也常存在格式错误的现象。

(2) 表身。表的主体部分,表身中常见的错误有数据错误,包括数值计算错误,比如平均值、总平均值等统计数据会因作者的马虎导致计算结果不正确;再就是对大量数据进行排序或分类时的纰漏,横列和纵列的内容调换位置时数据的输入错误;数字带有单位或百分号(应放在项目栏),单位重复(所有栏目单位相同时,应将单位标注在顶线的右侧,并不出现"单位"字样);表身中存在的空白栏,到底是代表这组项目未经测算还是计算结果为 0,要区分开来;数据中出现多位数时,没有在千分位空格;小数点没有对齐,显得整个表的数据杂乱,也不好核对,小数点前缺少"0"的情况也存在,这些都未按照标准规范化地书写。

(3) 表格线。三线表一般只包含一条顶线、一条栏目线和一条底线,在需要时会加入一些辅助线,用以解决栏目多层次问题。在添加辅助线时往往存在涵盖项目不清的情况,辅助线应根据表身内容,该长则长,该短则短,否则会造成内容表达上的歧义[23]。

(4) 表注。这一项并非所有表格都有,但需要备注的不能遗漏。经常看到一些表格中有数字加"﹡"或者加粗的,但并不知道这些特别的重点符号有何意义。正文中也没有交代,就一定要用表注进行说明。

图表的错误使得读者无法获得正确的信息,会影响期刊的整体质量水平。作为编辑,必须高度重视图表的加工和校对,力求在最后出版时,将正确的图表以最完美的方式呈现给读者[16,24]。

7.6 英文图表题名常用句型

7.6.1 关于"比较""对比""与……相比"句型的翻译

图表中关于比较类的题名很多,中心词为 comparison,主要句型有以下几种:①comparison between A and B;②comparison of A and B;③comparison of A with B;④comparison among/of A(s);⑤comparison of A to B。其中句型⑤在科技论文表达中不常用[25],主要含义为"比拟、把……比作"等,如 His comparison of the heart to a pump helped the children to understand its action(他把心脏比作泵有助于孩子们理解心脏的功能)。其他四种句型使用很普遍,举例如下。

(1) comparison between A and B。这种用法最常见,主要用于两者之间的比较,关键在于"between… and…"结构的搭配。其常见的变形结构有 comparison of C between A and B,如 comparison of yield between treatment 1 and 2(处理 1 与处理 2 的产量比较)。

(2) comparison of A and B。例如,"Figure 16 Comparison of calculated and measured

stress wave curves of pile B"[28]。

(3) comparison of A with B。例如,"Figure 2　Comparison of calculated dynamic velocity responses with measurement for case $H_s=0$"[29]。

(4) comparison among/of A(s)。例如,"Figure 6　Comparison of average degree of consolidation predicted from the proposed analytical solution and observations"[30]。

7.6.2　关于"影响""效应""作用""A对B的影响"等的常用句型

表达"影响""效应""作用""A对B的影响"等含义的名词中心词有effect(s)、influence(s)、impact(s)、role(s)等,常用的句型有:effect(s) of A on/upon B、influence of A on B、impact(s) of A on B、roles of A on B等。例如,"Figure 8　Influence of loading rate on the excess pore water pressure"[27],"Figure 10　Effect of soil plug density on APVPSP"[29]。

7.6.3　表示"相互关系""联系"等的常用句型

表示"相互关系""联系"等含义的常用名词中心词有relationship(s)、correlation(s)、relation(s)等。常用句型有以下几种。

(1) relationship(s)/correlation between A and B。例如,"Figure 2　Relationship between loading scheme and time"[27]。

(2) relation between A and B。例如,"Figure 12　The relation between the consolidation degree defined by settlement and pore water pressure"[30]。

(3) A and B。例如,"Figure 4　Experimental and theoretical valus"[29]。

(4) A - B。例如,"Figure 9　Void ratio-consolidation pressure relationship for different permeability coefficients"[30]。

(5) A versus/vs B。例如,"Figure 5　Annual average observed versus simulated water table depths from the surface"[27]。

(6) regression lines/curves of A and B,Fitted lines/curves of A and B。例如,"Figure 7　Regression lines of the linearized drop-size and velocity correlation equation"[28]。

7.6.4　关于各种"图"的英文表达

科技论文中的常用的插图(及其英文中心词)有简图(schematic(s))、略图(diagram)、示意图(schematic diagram)、流程图(flow chart)、直方图(histogram)、曲线图(curves)、分布图(distribution)、剖视图(cross-section、section)、三维图(three-dimension graph)、布局图(layout,arrangement)、结构图(structure、set-up)、轮廓图(contour)、等高线图(contour diagram)、影像图(image)、遥感图(remote sensing map)、地图(map)等。

(1)程序流程图、工艺流程图(flow chart)。例如,"Figure 5 Flow chart of calculation procedure of the horizontal bearing capacity analysis method for large diameter pile base based on modified strain wedge model"[31]。

(2)简图[schematic(s)]、略图(diagram)、示意图(schematic diagram)。例如,"Figure 4 Schematic of soil-pile interactionmodel"[29],"Figure 1 Diagram of pile-soil dynamic interaction model"[28],"Figure 1 Schematic diagram of soil"[32]。

(3)直方图(histogram)。例如,"Figure 5 Histogram of observed and simulated relatie solute concentrations in conventionally tilled field plot at 75cm depth,one day after a 5cm/h, 1-hduration rainstorm with tracer"[25]。

(4)布局图(layout、arrangement、plan)。例如,"Figure 1 Layout of the experimental field plots"[25]。

(5)剖视图(cross-section、section)、三视图(plan view、top view、right view)。例如, "Figure 2 Cross section of condenser"[25],"Figure 7 Top view of plow bottom and soil meshing with 35 and 45 angle of approach(s) to demonstrate thegreater soil volume meshed at $s=35$,and the soil volume not considered in the simulations"[25]。

(6)轮廓图(contour)、等高线图(contour diagram)。例如,"Figure 1 The contour diagram of extrudate's radial expansion ration"[25]。

(7)地图(map)。例如,"Figure 1 Sub-watershed map of the Leon River watershed"[25]。

(8)影像图(image)。例如,"Figure 1 Color infrared image of a research farm 17 days after planting:lighter color indicates soil with comparatively less amount of clay"[25]。

(9)分布图(distribution)。例如,"Figure 6 Distribution curves of the excess pore water pressure in the layered soils for different interface parameters"[27]。

(10)曲线图(curves)、照片(photo、photograph)。例如,"Figure 2 Photographs of production process using sludge"[25]。

7.6.5　表示"变化""改变"的常用句型

常用句型有:variation(s)of … with/to…;change(s)of … with…等。例如,"Figure 4 Variation of mean depth to the canal width"[25],"Figure 4 Variation of elastic modulus with confining pressure"[25]。

插图和表格是科技论文的重要表述手段,应当高度重视其制作和使用。现代科技书刊中的插图和表格多采用计算机绘制,计算机绘图软件种类丰富且功能齐全,科技工作者应当掌握必要的计算机绘图技术。不同专业领域的研究内容不同,其论文表述形式和采用的图表类型虽不尽相同,但都应遵循写实性、自明性、规范性和示意性的原则,精心设计制作和选用。只要以严谨的科学态度、精益求精的科研作风,掌握和运用先进的制作方法,就能够制作出规范、精美的论文插图和表格。

本章参考文献

[1]陈浩元.科技书刊标准化18讲[M].北京:北京师范大学出版社,1998:117-139.

[2]中国科学技术期刊编辑学会.科学技术期刊编辑教程[M].北京:人民军医出版社,2007:150-160.

[3]朱平盛,冯晓云.科技论文中图表的制作与使用[J].山东气象,2006,26(1):47-52.

[4]詹道友.医学论文图表的整理与编排[J].湖南医学,2000,17(2):110-112.

[5]庄庆德.科技论文撰写系列讲座(五)[J].名家讲坛,2008,27(6):1-3.

[6]刘万才,张凯,龚玉琴,等.科技论文撰稿中图表的规范应用[J].中国农技推广,2007,23(9):42-44.

[7]方志荣.图表制作的标准化方法[J].石油工业技术监督,2004,20(9):42-43.

[8]陈淑珍.图表在科技期刊中的表达[J].沈阳工业学院学报,2003,22(2):93-94.

[9]全国信息与文献标准化技术委员会.科技报告编写:GB/T 7713.3—2014[S].北京:中国标准出版社,2014.

[10]全国标准化原理与方法标准化技术委员会.标准化工作导则 第1部分:标准化文件的结构和起草规则.GB/T 1.1—2020[S].北京:中国标准出版社,2020.

[11]吴重龙,白来勤.编辑工作手册[M].北京:华艺出版社,2004.

[12]朱平盛.科技论文写作要点[J].山东气象,2005,25(4):8-12.

[13]全国量和单位标准化技术委员会.国际单位制及其应用:GB 3100—1993[S].北京:中国标准出版社,1993.

[14]全国量和单位标准化技术委员会.有关量、单位和符号的一般原则:GB 3101—1993[S].北京:中国标准出版社,1993.

[15]王立名.科学技术期刊编辑教程[M].北京:人民军医出版社,1997.

[16]熊英,欧阳贱华,於秀芝,等.科技论文中图表的加工和校对[J].编辑学报,2011,23(2):123-125.

[17]陈玲.科技论文图标的编辑加工[J].泸天化科技,1995,3:236-239.

[18]吕广玉,贺六连,曾湘华,等.石油科技论文的图、表规范表达[J].大庆石油学院学报,1996,20(1):129-132.

[19]刘瑜君,李玉江.论科技论文中的图表编校[J].湖北师范学院学报(自然科学版),2016,36(3):255-257.

[20]石幸利.科技期刊中图表及公式的编排规范[J].重庆科技学院学报(自然科学版),2013,15(3):174-176.

[21]张冬冬,赵春秀,陈欣,等.科技论文图表的编辑加工[J].鞍山师范学院学报,2015,17(6):106-108.

[22]窦春蕊,赵粉侠,马勤,等.科技论文作者应掌握的国家标准与规范(四):正文中插图和表格的规范化[J].陕西林业科技,2001(4):62-64.

[23]刘祥娥,林琳.科技期刊三线表使用中的常见问题分析[J].中国科技期刊研究,2007,18(5):900-901.

[24]马智成,夏继军.科技期刊中图表的校对方法[J].编辑学报,2012,24(S1):24-25.

[25]王应宽,蒲应龚.科学学术论文英文图表题名译写举隅[J].编辑学报,2001,13(S1):24-26.

[26]刘浩,吴文兵,蒋国盛,等.土塞效应对管桩低应变测试波速的影响研究[J].岩土工程学报,2019,41(2):383-389.

[27]YANG X Y,ZONG M F,TIAN Y,et al. One-dimensional consolidation of layered soils under ramp load based on continuous drainage boundary[J]. International Journal for Numerical and Analytical Methods in Geomechanics,2021,45(6):738-751.

[28]LIU H,WU W B,NI X Y,et al. Influence of soil mass on the vertical dynamic characteristics of pipepiles[J]. Computers and Geotechnics,2020,126[2020-09-20]. http://doi.org/10.1016/j.compgeo.2020.103730.

[29]WU W B,LIU H,YANG X Y,et al. New method to calculate apparent phase velocity of open-ended pipepile[J]. Canadian Geotechnical Journal,2020,57(1):127-138.

[30]TIAN Y,WU W B,JIANG G S,et al. Analytical solutions for vacuum preloading consolidation with prefabricated vertical drain based on elliptical cylinder model[J]. Computers and Geotechnics,2019,116[2020-09-20]. https://doi.org/10.1016/j.compgeo.2019.103202.

[31]邢康宇,吴文兵,张凯顺,等.基于改进应变楔模型的大直径桩基水平承载力分析方法[J].安全与环境工程,2020,27(3):200-207.

[32]TIAN Y,WU W B,JIANG G S,et al. One-dimensional consolidation of soil under multi-stage load based on continuous drainage boundary[J]. International Journal for Numerical and Analytical Methods in Geomechanics,2020,44:1170-1183.

第 8 章 论文表达与语言编辑

8.1 时态的运用

英文科技论文撰写过程中常用的时态有两种,即一般过去时和现在时。

通常情况下,科技论文中采用**一般过去时**来表述所做研究工作的时间相对于写作时间来说是过去的,尤其是一些实验(试验)研究工作。例如,"First, field tests were conducted to measure the vertical earth pressure on the TTI culvert. Then, numerical simulations were performed to investigate the performance of TTI culverts. With the validation of the numerical model, parametric studies were carried out to investigate the important influencing factors on the culvert performance…""The culvert was installed above a moderately weathered mudstone. A composite layer(1.0m in thickness) mixed with cement, sand and gravel was placed below the culvert to create a leveling ground."。

采用**一般现在时**来陈述客观事实,比如陈述研究目的、描述结果、得出结论、提出建议或讨论等。涉及公认事实、自然规律、永恒真理等,也要用**一般现在时**。例如,①"This study is performed to investigate the complex soil-structure interaction and to provide reference for the design of trapezoidal trench installation(TTI) culverts."(表述研究目的);②"Table 4 lists …"(列表描述结果);③"Figure 3 shows that …"(画图描述结果);④"This study reveals that the vertical earth pressure on the TTI culvert is significantly different from those on the EI and TI culverts. The magnitude of the vertical earth pressure and deformation of the TTI culvert are influenced by the soil arches as long as the backfill is high enough."(表述研究结论);⑤"Generally, it is suggested to consider the soil arch effects in the design of TTI culverts."(提出建议);⑥"It can be found that the vertical earth pressures on the culvert and on the adjacent backfill mass increase with the height of backfill."(描述公认事实)。

论文中描述研究领域内已存在的知识和结论时,通常采用**一般现在时**以表示对理论贡献者的尊重。因此,introduction、discussion 和 conclusion 中的大部分内容应该采用一般现在时。摘要(abstract)、方法(methods)和内容(content)中介绍自己的工作时,由于研究工作在过去完成,所以通常采用**一般过去时**。表 8-1 列出了英文科技论文主要部分使用时态的一般规则。

表 8-1 科技论文时态使用的一般规则

时态	Introduction	Methods	Results	Discussion	Conclusion
一般现在时	大量使用	很少使用	大量使用	大量使用	大量使用
一般过去时	很少使用	大量使用	很少使用	很少使用	很少使用

8.2 语态的运用

在早期的论文写作中，由于科技论文主要说明事实经过，至于研究工作是谁做的无需逐一说明，为强调动作承受者，主要采用**被动语态**。比如"The commercial FE software ABAQUS was adopted to simulate the foundation excavation process.""In this study, a full-scale experiment and FEM simulation were conducted to evaluate the variation of vertical earth pressures and soil arching in backfill and to examine the accuracy of the methods recommended by different design codes."[1]"The distribution and the dissipation of excess pore water pressure in the soil are examined."等。

21世纪以来，大量学者采用**主动语态**撰写英文科技论文。比如"This paper presents the failure of soil masses during shield tunneling.""This paper presents a numerical study on the performance of embankments on soft ground with three different reinforcement conditions…"。不难发现，主动语态往往比被动语态更加简洁明了、表达有力。不仅如此，人称代词we也时常出现在科技论文中。比如"We performed a biaxial test simulation to calibrate the contact parameters for the discrete element model."。

在科技论文写作中，主动语态和被动语态可以同时使用，也可以互相转换。具体使用哪种语态取决于句子所要强调的重点，同时应该注意表达的简练性和语义的精准性。在通常情况下，主动语态的表达更为准确，且易于阅读，因而目前大多数国际期刊提倡使用主动态，国际知名科技期刊 Nature、Cell 等尤其如此，其第一人称和主动语态的使用十分普遍。

8.3 表达方式和技巧

"严谨是科研的灵魂。"英语论文的写作既要精简，也要力求精确。合适的表达方式能够让读者迅速地理解作者的深刻要义，使读者快速进入作者的写作逻辑，从而跟上作者的研究思路，掌握作者的研究工作。一般说来，非英语母语作者存在以下几个方面的表达问题。

(1)句子内容不连贯。
(2)论述的逻辑性差。
(3)语法不准确。
(4)用词或者表达论点不准确。
(5)论文各部分逻辑性不强，结构混乱等。

一般说来，一个句子的词汇量大概在 12～17 个单词之间。每个句子表达一个相对完整的意思或思想（idea）。

段落的叙述要遵循一定的逻辑主线，每段的第一句话扼要概述该段所要陈述的内容，每段的主题内容应明确、单一。在写作过程中，大致包含以下方面：

（1）使用大致包括 IMRDC（introduction、methods、results、discussion、conclusion）结构的论文写作模式。

（2）使用简短的句子，用词应为潜在读者所熟悉的专业术语。

（3）注意表述的逻辑性，尽量使用指示性的词语来表达论文的不同部分（层次）。例如使用"研究表明……"（"We find that …""It is found that …""This research reveals that …""The numerical results show that …"）表示结果；使用"通过对……的分析，……认为……"（"However, this finding was not always evident from the study discussed in this section.""Based on …, we suggest that …"）表示讨论等。

英文论文的基本框架大体包括 title、introduction、methods、results、tables and figures、discussion、conclusion、references 等部分。各个部分的具体要求如下。

（1）**title**：be short, accurate and unambiguous; give your paper a distinct personality。

（2）**introduction**：what is known; what is unknown; why we did this study。

（3）**methods**：participants, subjects; measurements; outcomes and explanatory variables; statistical methods。

（4）**results**：sample characteristics; univariate analyses; bivariate analyses; multivariate analyses; objective performance。

（5）**tables and figures**：no more than six tables and more than ten figures; use Table 1 for sample characteristics; put most important findings in a figure。

（6）**discussion**：state what you have found; outline the strengths and limitations of the study; discuss the relevance to current literature; outline your implications with a clear "so what" and "where now"。

（7）**conclusion**：to show a new idea; to refine the findings; to be general applicable; to be succinct generalization。

（8）**references**：all citations must be accurate; include only the most important, most rigorous, and most recent literature; quote only published journal articles or books; never quote "second hand"; cite only 20 - 35 references。

1. 引言

引言部分一般要对已有研究工作进行综述，指出已有的研究工作所取得的成就及有待改进的地方，从而引出文章的创新点及研究思路。常用的表述方式包括：

（1）回顾研究背景。常用词汇有 review、summarize、present、outline、describe/consider 等。常见表达有：

One of the outstanding contributions of …

Succeeding sb's study…

In summary, the literature **reviewed** above indicates that…

Sb. **summarized** a large number of solutions obtained for an ×× problem and presented a closed-form solution for sth.

Most studies related to "the issue" have **considered** the case of ××.

Sb. **presents** an analytical solution for ××.

Minor modifications to the test as **outlined** by sb. enable quantitative assessment of ××.

（2）说明写作目的。常用词汇有 purpose、is performed to、attempt、aim、goal、present、focus on、concentrate on、objective、address 等。常见的表达有：

For ××, preliminary methods to assess tunneling-induced damage have been **proposed**, including the limiting tensile strain method.

For this **purpose**, the effects of ×× were examined.

This study **is performed to** investigate ××.

This study **attempted** to ××.

The present study is **aimed** at demonstrating the feasibility of the proposed methods.

This paper **presents** results from a program of geotechnical centrifuge tests that aim to investigate the tunnel-pile-structure interaction problem.[2]

The work presented in this paper **focuses on** ××.

The main **objective** of this study was to propose ××.

This paper **addresses** the problem of tunnel-pile-structure interaction and considers ××.

（3）介绍论文的重点内容或研究范围。常用词汇有 although/however、present、include、focus、study、emphasize、attention、introduce 等。常见的表达有：

Sth. is believed to benefit from sth. because it increases…, **however**, this finding was not always evident from the study discussed in this section.

Although the predominant periods of S waves and surface waves are almost identical, the vertical amplitude distributions of these waves can be very different in surface layers at a soft-soil site.

A baseline case was selected, which **included** ××.

Results are then presented that **focus on** soil and structural deformations.

A parametric **study** was numerically carried out to investigate the important influencing factors on the performance of ××.

Finally, the enhanced BBM is **introduced** and its comparison against the experimental results is **analyzed**.[3]

2. 方法

方法部分是论文的核心部分，主要介绍作者开展研究工作的过程。常用的表述方式包括：

(1) 介绍研究、试验或者数值模拟过程。常用词汇有 test、examine、investigate、discuss、consider、simulate、model 等。常见的表达有：

The **test** is performed in two modes, A and B.

At the same time, the mixture was also poured into the PVC column mold with a diameter of 38 mm and a height of 76 mm to **examine** ××.

For this reason, "××" were also conducted to **investigate** the effect of ××.

The influencing factors, such as ×× and ××, on the interaction of ×× systems are **investigated** and **discussed**.

In this study, different ×× and ×× are **considered**. The time dependent behavior of ×× should **be considered** in the design of ××.

A schematic of the apparatus is shown in Fig. 1, which was originally developed **to simulate** ××; A numerical model based on a ×× program has been established **to simulate** the performance of ××. The three different cases **are simulated** by a numerical method to analyze ××.

The "reinforced concrete box culverts" were **modeled** using ××.

(2) 说明研究、试验或数值模拟方法。常用词汇有 measure、estimate、calculate、method、software 等。常见的表达有：

The **measured** vertical earth pressures from the experiment were compared with those from the current theoretical methods; Earth pressure cells were used **to measure** the pressures on the culvert due to the backfill.

Sb. proposed a theoretical method to **estimate** the load acting on "××". Using the deep structural model, input wave fields **were estimated**.

In addition, a new simplified theoretical **method** was deduced **to calculate** the vertical earth pressure on the ITI box culvert[4]. The average strength **is calculated** using Eq. ×.

A numerical model using **Plaxis software** is employed to investigate ××.

(3) 介绍应用、用途。常用词汇有 use、apply、adopt、utilize、employ、perform、carry out、conduct、establish 等。常见的表达有：

A soft layer of EPS board with 30 mm in thickness setting above a box culvert was **used** as a load reduction material; Various techniques may be **used** to address this problem, such as ××.

Toinvestigate the friction effect between soil and culvert, a 5-mm-thick shotcrete was evenly **applied** on both sidewalls.

Ground treatments are often **adopted** to enhance the ground bearing capacity.

A numerical method was **adopted** to conduct a parametric analysis.

The layer-wise summation method is **utilized** to calculate the land subsidence caused by dewatering.

A finite element program has been **employed** to investigate ××.

Then, numerical simulations were **performed** to investigate the stress states of ××.[5]

Model test and numerical simulation were **carried out** to investigate the performance of EI and ITI box culverts.

A parametric study was **carried out** to investigate the effect of influencing factors.[6]

In addition to the above methods, nonlinear regression analysis can be **conducted** to obtain a regression equation for determining the vertical earth pressure on the culvert.

A maximum-backfill-height model was **established**, then each element layer was activated based on the layered filling steps.

3. 结果

结果部分主要是对试验、数值模拟和理论推导所获得的结果进行展示,常用的表述方式包括:

(1)展示研究结果。常用词汇有 show、result、present、depict、draw、list、give、summarize 等。常见的表达有:

Fig. × **shows** the variation of ××.

The **results demonstrate** that the vertical earth pressure concentrated on the edge of the top slab.

The shear stress between the surrounding soil prism and the central soil prism **results in** backfill pressure concentration on the culvert.

Fig. A **presents** the variations of ××.

Figs. A and B **depict** the development of ××.

The following conclusions can **be drawn**.

The measured final settlements for all test conditions are **listed** in Table A.

The field test results of ×× are **given** in Fig. A.

The stress-strain parameters derived from the results of the triaxial tests are **summarized** in Table A.

(2)介绍结论。常用词汇有 summary、conclude、drawn 等。常见的表达有:

The performance of the ×× is **summarized** as follows.

The research analyzed the vertical loads on concrete box culverts under high embankments based on a field test and **concluded** that the vertical load on the structure was independent of the ratio of the height of fill to the width of the structure.[7]

It can be **concluded** that the vertical earth pressure on the culvert installed in the trench is not always lower than that calculated by the linear earth pressure theory based on the overburden stresses.

The conclusions can be **drawn** as follows. The following conclusions may be **drawn** from this study.

4. 讨论

讨论部分主要针对论文研究工作所获得的结果、结果的适用范围,以及结果的应用价值等进行详细的讨论。常用的表述方式包括:

(1)讨论论文的主要结论和作者的主要观点。常用词汇有 suggest、expect、describe 等。常见的表达有:

Overall, when considering similar budgets, a wide and shallow reinforced area in the subgrade layer with proper stiffness is **suggested** instead of a narrow and deep reinforcement with excessive stiffness.

This **suggests** that the drained behavior is more likely to give rise to issues associated with large displacement, rather than complete collapse.

Although this method reduces the vertical earth pressure on the top slab, it has insignificant effect on the reduction of the foundation contact pressure as **expected**.

As **described** in Section 4.2.1, the deformation restraint effect of soil movement inside the pit and the deformation compatibility arising in the soils and at the soil-wall interface can account for the computed wall behavior at deep locations.

(2)说明论证,数据支撑结果或规律方面。常用词汇有 provide、indicate、demonstrate、confirm、explain、reveal 等。常见的表达有:

Horizontal geosynthetic reinforcement combined with vertical reinforcement using piles or pile walls can **provide** an economical and effective solution to ground improvement to support high embankments.[8]

The numerical results **indicate** that the tension along the geosynthetic layer is not uniform, it decreases with the distance from the centerline of the embankment.

Due to the high degree of fracturing in relation to the profile scale, the failure surfaces described by the shear strain accumulation zones **indicated** approximately circular surfaces which followed lines of least resistance.

The results of arandom field survey of 102 HRC culverts **demonstrated** that structural problems occur frequently in culverts.

Field tests were conducted to measure the settlement ratioand it **demonstrated** that the magnitude and direction of the movement of the central soil prism on the structure are related to the surrounding soil prisms.

The frictionon ITI box culvert sidewall is significantly larger than that of EI box culvert. It **confirms** that the additional vertical earth pressure on adjacent backfill mass leads to an increase in friction on ITI box culvert sidewall.

The effect of the exterior arch is more significant when the height of the backfill increases. This **explains** the decrease of the vertical earth pressure coefficient when the backfill is high enough.

Results of the field test and numerical simulation **reveals** that the average foundation pressure increased with an increase in backfill height.

(3)讨论部分经常会根据发现的结果或规律,对其应用方面给予推荐或建议。常用词汇有 suggest、recommend、propose、necessary、expect 等。常见的表达有:

Generally, it is **suggested** to consider the soil arch effects in the design of TTI culverts.

The geosynthetics combined with pile walls is **recommended** in the design of soft ground improvement to control the lateral movement and the total and post-construction settlements.

In engineering practice, a large slope angle and/or a small trench bottom width is **recommended** to minimize the vertical earth pressure on the culvert.

A culvert-soilinteraction model was therefore **proposed** to consider the friction on the culvert sidewall.

It is therefore **necessary** that the engineers consider carefully the inherent limitations that a simplification from 3D to a 2D analysis is entailed.

It is **expected** that an improvement of the culvert foundation would increase the stiffness of foundation thus inducing stress concentration and increasing the pressure on the top of the culvert.

此外,研究工作一般都是在科研项目的支持下进行的,在论文的结尾往往会列出科研项目的名称和编号表示感谢。致谢(acknowledgments)的常见格式如下。

This work is supported by the National Natural Science Foundation of China(NSFC) (No. ××). The authors would like to express their appreciation to this financial assistance. The authors would also like to thank Prof. ×× at the University of ×× for useful discussions, which helped to improve the quality of this paper.

8.4 如何使用地道的英语表达方式

英文论文写作过程中,母语非英语的学者,很难像英语为母语的研究者那样流畅地表达自己的思想和研究成果。再加上中西方文化的差异,造成许多表达方式的差异,或者说英语表达不地道。解决这一问题的思路之一就是多读高水平的国际文献,尤其是文献作者的母语为英语的高水平期刊论文,熟练掌握论文中描述专业问题的特定的表达方式和表达习惯,然后在英文论文的写作过程中合理运用这些表达方式。

1. 连接词

科技论文中连接词的使用可以使得文章更为紧凑,启承转接更加流畅自然,逻辑性更强。常见的连接词用法如下。

(1)表示转折或重点强调某一方面。常用的连接词有 however、but、while、although、rather than、in contrast、by contrast 等。典型例子如下:

It is typical that the modulus of soil depends on the confining stress. In this study, **however**, a constant value was used for the modulus of backfill for simplification since the modulus of compacted backfill is less dependent on the confining stress.

This research reveals that soil arch formed when the backfill on the culvert reached a certain height, **but** it was unstable. [9]

The effective normal stress above the geosynthetic increases rapidly **while** the effective normal stress below the geosynthetic decreases rapidly and reaches to a small value.

The ITI method cannot reduce the FCP, **although** it can reduce the VEP on the culvert top slab.

The parametric study mainly focuses on the construction period **rather than** the post construction period. The ground soil consolidation in the post construction period may influence the stress state of the RC culvert.

Damage to HRC culverts are rarely caused by subgrade layer failure beneath the culvert foundation. **In contrast**, if the ground bearing capacity of an HRC culvert is underestimated, the adoption of an improper ground treatment method to reinforce the subgrade layer may result in structural damage.

The TI method is often costly for most projects, **by contrast**, the ITI methods are likely to provide an economical and effective solution to reduce the vertical load on the top of box culvert.

(2) 表示并列或补充说明。常用的连接词有 also、as well as、in addition、moreover、furthermore 等。典型例子如下：

These coefficients **also** fluctuated **but** opposite to those on the culvert; ⋯, it is **also** shown that ××.

The main purpose of such a study is to capture the effects of the tensile stiffness of geosynthetics, the number of geosynthetic layers, the elastic modulus of pile walls, **as well as** pile wall distance ratio on the structural performance of geosynthetics and pile walls reinforced embankment systems.

In general, the tension in the geosynthetic layer decreases with the distance from the centerline of the embankment. **In addition**, there is an obvious increase in the tension over the pile walls. **Moreover**, the maximum tension in the geosynthetics layer approximately occurs below the shoulder of the embankment.

Furthermore, the tensile stiffness of the geosynthetics has a pronounced influence on the horizontal displacement at the toe and the side pile wall. **However**, the influence on the total settlement of the embankment and the differential settlement at the ground surface elevation is small.

(3) 表示原因或由此导致的结果。常用的连接词有 because、due to、consequently、as a consequence、as a result、thus、therefore 等。典型例子如下：

The stress state of the culvert in the ground is different from that of a footing in the soil **because** the interaction of the culvert and the soil mainly depends on the characteristics of the embankment fill and the foundation soil.

The arch springings are exactly located on the corners of the top slab. **Consequently**, the vertical earth pressure on the center of the top slab is gradually transferred to the corners of the top slab.

The deformation in the compressible inclusion provides mobilization of the shear strength of the soil above the culvert, and **as a consequence** the vertical earth pressure remains lower than the overburden pressure caused by the weight of the soil.

The difference in stiffnessinduces the settlement of M_{21} and M_{23} being greater than that of M_{22}. **As a result**, differential settlement also exists between the inner and outer soil columns above the culvert top plane, which induces partial overburden pressure of outer columns adding to the inner soil column through lagging effect. **Thus**, lead to the increase of VEP on the top slab of EI box culvert.

With an increase of the height of the backfill, a new soil arch formed and thecoefficient of the vertical earth pressure on the culvert increased again. **However**, this phenomenon could not be simulated by the numerical method, **therefore**, some differences exist between the numerical results and the field data.

（4）表示研究发现的结果与认识相同或相左。常用的连接词有 unlikely、similarly 等。典型例子如下：

The differential pressure between average foundation and embankment pressuresis less than 10%. Thus, it is highly **unlikely** that failure of the subgrade layer beneath the culvert foundation will occur if the subgrade layer can bear the dead weight of the fill mass adjacent to the culvert.

The vertical earth pressure concentrates on the culvert due to the shear stresses between the central backfill mass over the culvert and the surrounding backfill masses. **Similarly**, the slopes of the trench resist the descent of the backfill masses between the culvert and the slopes.

（5）其他常用的连接词有 first(second、third)、finally、for example、meanwhile、in fact、because、in order to 等。典型例子如下：

First, in the linear soil response analysis of the soil response study, a wave field was reproduced well as the simulated soil responses based on the wave field used as input… **Second**, in the nonlinear soil response analysis in which the onlinear parameters of Table 2 were used… **Third**, in the liquefaction soil response analysis, only the surface ayers above 10 m were supposed to have liquefaction potential…

Finally, a specific and quick procedure for the evaluation of the mechanical behavior of the support of a base tunnel has been illustrated.

For the design of concrete box culverts installed in trenches, the general concept is to consider that the vertical earth pressure on the top of the culvert is less than that computed using the linear earth pressure theory, **for example**, the formula based on the Chinese General Code for Design of Highway Bridges and Culverts.

The excess pore water pressure is dissipating, **meanwhile**, the effective stress is increasing, during the construction of the embankment.

This coefficient is defined as the ratio of the vertical earth pressure at the top level of the culvert to the overburden pressure of the backfill at the same location. **In fact**, the vertical earth pressure depends on the height of the backfill, the dimensions of the trench and the culvert, the properties of the backfill and foundation soil, etc.

In order to reduce the backfill load on the culvert, several methods are commonly used in construction process, among which, the imperfect-trench-installation box culvert(IBC)is usually used in the embankment engineering.

2. 常用词汇

英语科技论文写作中的常见词汇及其特殊用法如下。

(1)research、study、investigate。research 可作名词、动词,但通常作名词用,很少见到动词的用法。当句中需要动词时,常用 study 或 investigate。

(2)detail、detailed。

The properties of this compound were studied in detail.

The detailed properties of this compound were studied.

The details of the properties of this compound were studied.

(3)follows、following。

The results are as follows:…

We got the following results:…

(4)increase、decrease。均可作名词和动词。

We can observe an increase in the reaction rate.

The reaction rate increases.

(5)focus、concentrate。focus 可作名词和动词,concentrate 作动词。

The focus of this paper is …

This paper focuses on …

Our study focuses on …

We focus our study on …

Our study is focused on …

We concentrate our study on …

Our study is concentrated on …

(6)effect、affect。effect 作名词,affect 作动词。

(7)compose、consist。

A is composed of B and C.

A consists of B and C.

(8)increase、improve。increase 主要指数值上的增加，improve 主要表示性质方面的增加或改善。

3. 常见表述

在阅读文献时，注意积累一些常见的表达方式，有利于在英文论文写作中应用。本书仅列出以下几个例子。

(1)做……有意义：It makes sense to…。

(2)对……投入了很多的关注：Much attention has been paid to…。

(3)对……做出了很大的努力：Great efforts have been made to…。

(4)值得提及……：It worth mentioning/noting that…。

(5)考虑到……：Account for/Consider…。

(6)不失一般性地：without loss of generality。

(7)……是毫无疑问的：There is no doubt that…/Undoubtedly，…。

(8)……已经被广泛地应用于……：…has been widely used/adopted to/applied in…。

(9)关于……：With regard of /Regarding…。

(10)在……领域，已做了大量工作：Extensive work has been done in the area of…。

(11)……已经得到公认：It is well known/accepted that…。

(12)……与相似……：…in analogy with/similar to…。

(13)与……一致：be consistent with…、be in agreement with…、be the same as…、be in common with…。

(14)……起主要角色：…plays the main role。

(15)致力于……：be devoted to…。

(16)对……敏感：be sensitive to…。

(17)由……决定：be determined by…。

(18)定义为……：be defined as…，be referred to…。

(19)在……背景下：in the context of…。

(20)做……非常有趣：It is interesting to…。

8.5 实例介绍

为了进一步阐述英语科技论文写作中语言的表达和编辑，本节以被引用上千次的计算力学经典文献 *Isogeometric analysis：CAD, finite elements, NURBS, exact geometry and mesh refinement*[10]中的引言为例，介绍其英语表达方式。该文献的"Introduction"如下，请注意文中粗斜体的典型表达。

In this paper we introduce a new method for the analysis of problems governed by partial differential equations such as, for example, solids, structures and fluids. The method has many features **in common with** the finite element method and some features in common with meshless methods. **However**, it is more geometrically based and takes inspiration from Computer Aided Design(CAD). A primary goal is to be geometrically exact **no matter how** coarse the discretization. Another goal is to simplify mesh refinement by eliminating the need for communication with the CAD geometry once the initial mesh is constructed. Yet another goal is to more tightly weave the mesh generation process within CAD. In this work we introduce ideas in pursuit of these goals.

It is interesting to note that finite element analysis in engineering had its origins in the 1950s and 1960s. Aerospace engineering was the focal point of activity during this time. By the late 1960s the first commercial computer programs (ASKA, NASTRAN, Stardyne, etc.) appeared. Subsequently, the finite element method spread to other engineering and scientific disciplines, and now its use is widespread and many commercial programs are available. **Despite the fact that** geometry is the underpinning of analysis, CAD, as we know it today, had its origins later, in the 1970s and 1980s. A highly-recommended introductory book, with historical insights, is Rogers[1]. This perhaps explains why the geometric representations in finite element analysis and CAD are so different. Major finite element programs were technically mature long before modern CAD **was widely adopted**. Presently, CAD is a much bigger industry than analysis. Analysis is usually referred to as Computer Aided Engineering(CAE) in market research. It is difficult to precisely quantify the size of the CAE and CAD industries but current estimates are that CAE is in the \$1-\$2 billion range and CAD is in the \$5-\$10 billion range. The typical situation in engineering practice is that designs are encapsulated in CAD systems and meshes are generated from CAD data. **This amounts to** adopting a totally different geometric

这篇论文的引言一气呵成，逻辑性非常强，用词非常地道。

第一段就讲清了文章的目的和论文的创新点：几何精确、简化网格细化过程和将网格与CAD关联。

第二段介绍有限元与CAD的发展历史，包括区别与联系。

description for analysis and one that is only approximate. ***In some instances***, mesh generation can be done automatically but in most circumstances it can be done at best semi-automatically. ***There are still situations in major industries in which*** drawings are made of CAD designs and meshes are built from them. It is estimated that about 80% of overall analysis time is devoted to mesh generation in the automotive, aerospace, and ship building industries. In the automotive industry, a mesh for an entire vehicle takes about four months to create. Design changes are made on a daily basis, limiting the utility of analysis in design if new meshes cannot be generated within that time frame. Once a mesh is constructed, refinement requires communication with the CAD system during each refinement iteration. This link is often unavailable, ***which perhaps explains why*** adaptive refinement is still primarily an academic endeavor rather than an industrial technology.

The geometric approximation inherent in the mesh can ***lead to*** accuracy problems. One example of this is in thin shell analysis, which is notoriously sensitive to geometric imperfections; see Fig. 1. The sensitivity to imperfections is shown in Fig. 1b in which the buckling load of a geometrically perfect cylindrical shell is compared with shells in which geometric imperfections are introduced with magnitudes of 1%, 10%, and 50% of the thickness. ***As may be seen, there is a very considerable*** reduction in buckling load with increased imperfection.

Euler solvers in the 1980s and 1990s. The problem and its solution were identified in the thesis of Barth[2]. Piecewise linear approximations of geometry were the root cause. Smooth geometry completely eliminated the entropy layers even when the flow fields were approximated by linear elements on the curved geometry; see Fig. 2. This result explains why methods which employ smooth geometric mappings are widely used in airfoil analysis (see[3]). ***It is also well known in*** computational fluid dynamics that good quality boundary layer meshes significantly improve the accuracy of computed

第三~第五段介绍了几何精确在固体和流体力学计算中的重要性。

wall quantities, such as pressure, friction coefficient, and heat flux; see Fig. 3.

Sensitivity to geometry has also been noted in fluid mechanics. Spurious entropylayers about aerodynamic shapes were the bane of compressible. The construction of finite element geometry (i.e., the mesh) is costly, time consuming and creates inaccuracies. ***It is clear from*** the smaller size of the CAE industry compared with the CAD industry that the most fruitful direction would be to attempt to change, or replace, finite element analysis with something more CAD-like. This direction was taken in the development of the RASNA program Mechanica, in which exact geometry ***in conjunction with*** a p-adaptive finite element procedure was utilized. However, the lack of satisfaction of the isoparametric concept led to theoretical questions which were addressed in later versions of the code by abandoning the exact geometry in favor of high-order polynomial approximations[7]. The use of a fixed polynomial approximation to geometry has been shown by Szabo et al.[8] to be limiting. As solution polynomial order is increased, the error plateaus at some level and cannot be further reduced (see Fig. 4). The seriousness of this result ***is compounded by the fact that*** computed quantities defined on boundaries are usually the most important ones in engineering applications, and this is where geometric errors are most harmful. Furthermore, most finite element analyses are still performed with low-order elements for which geometric errors are largest. The success of RASNA, which was later acquired by Parametric Technology Corporation(PTC), a CAD company, was due to its tight linkage with CAD geometry and, perhaps more importantly, its consequent ability to provide adaptive p-refinement and thus more reliable results. The present methodology is similarly inspired, but attempts to more faithfully adhere to CAD geometry and eliminate the finite element polynomial description entirely. (The p-method is de-scribed in Szabo and Babuska[9] and Szabo et al.[8].)

The approach we have developed is based on NURBS (Non-Uniform Rational B-Splines), a standard technology employed in CAD systems. We propose to match the exact CAD geometry by NURBS surfaces, then construct a coarse mesh of "NURBS elements". These would be solid elements in three-dimensions that exactly represent the geometry. **This is obviously not** a trivial task and one that deserves much study **but** when it can be accomplished it opens a door to powerful applications. Subsequent refinement does not require any further communication with the CAD system and is so simple that it may facilitate more widespread adoption of this technology in industry.

There are analogues of $h-$, $p-$, and $hp-$ refinement strategies, and a new, higher-order methodology emerges, $k-$refinement, which seems to **have advantages of** effciency and robustness over traditional $p-$refinement. All subsequent meshes retain exact geometry. Throughout, the isoparametric philosophy is invoked, that is, the solution space for dependent variables is represented in terms of the same functions which represent the geometry. **For this reason**, we have dubbed the methodology isogeometric analysis.

NURBS are not a requisite ingredient in isogeometric analysis. We might envision developing isogeometric procedures based on" A-patches" (see[11-16]) or " subdivision surfaces" (see[17-19]). **However**, NURBS seem to be the most thoroughly developed CAD technology and the one in most widespread use.

The body of this paper begins with a tutorial on B-splines (B-splines are the progenitors of NURBS), followed by one on NURBS. We then describe an analysis framework based on NURBS. This is followed by sample applications in linear solid and structural mechanics and some introductory calculations in fluids, namely, ones involving classical test cases for the advection-diffusion equation. Various refinement strategies are studied and, in cases for which exact elasticity solutions are available, optimal rates of convergence are attained. The structural problems include some applications to thin shells modeled as solids. **The approach is seen to** handle these situations

第六、第七段介绍了该文的研究内容。第八段概述了论文的结构组成。

remarkably well. In the fluid calculations, we employ the SUPG formulation and consider difficult test cases involving internal and boundary layers. We observe that, by employing high-order, *k* - refinement strategies, convergence toward monotone solutions is obtained. ***This surprising result seems to*** contradict numerical analysis intuitions and suggests the possibility of linear difference methods that are simultaneously robust and highly accurate. We close with conclusions and suggestions for future work.

本章参考文献

[1] CHEN B G, ZHENG J J, HAN J. Experimental study and numerical simulation on concrete box culverts in trenches[J]. Journal of Performance of Constructed Facilities, 2010, 24(3):223-234.

[2] FRANZA A M, MARSHALL A M. Centrifuge modeling study of the response of piled structures to tunneling [J/OL]. Journal of Geotechnical & Geoenvironmental Engineering, 2018, 144(2)(2019-02-20)[2020-11-20]. https://doi.org/10.1061/(ASCE)GT.1943—5006.0001751.

[3] ROMERO E, SANCHEZ M, GAI X, et al. Mechanical behavior of an unsaturated clayey silt: an experimental and constitutive modelling study[J]. Canadian Geotechnical Journal, 2019, 56(6):1461-1474.

[4] CHEN B G, MENG Q D, YAN T F, et al. New simplified method for analysis of earth pressure on the imperfect trench installation box culvert[J]. International Journal of Geomechanics, 2020, 20(10)[2020-12-23]. https://ascelibrary.org/doi/epdf/10.1061/%28ASCE%29GM.1943-5622.0001842. DOI:10.1061/(ASCE)GM.1943-5622.0001842.

[5] CHEN B G, SONG D B, MAO X Y, et al. Model test and numerical simulation on rigid load shedding culvert backfilled with sand[J]. Computers and Geotechnics, 2016, 79:31-40.

[6] CHEN B G, SUN L. Performance of a reinforced concrete box culvert installed in trapezoidal trenches[J]. Journal of Bridge Engineering, 2014, 19(1):120-130.

[7] BENNETT R M, WOOD S M, DRUMM E C, et al. Vertical loads on concrete box culverts under high embankments[J]. Journal of Bridge Engineering, 2005, 10(6):643-649.

[8] ZHENG J J, CHEN B G, LU Y E, et al. The performance of an embankment on soft ground reinforced with geosynthetics and pile walls[J]. Geosynthetics International, 2009, 16(3):173-182.

[9] CHEN B G, JIAO J J, SONG D B. Study of vertical earth pressure on reinforced concrete box culvert: trapezoidal trench installing culvert[J]. Rock and Soil Mechanics, 2013, 34(10): 2911-2918.

[10] HUGHES T J R, COTTRELL J A, BAZILEVS Y. Isogeometric analysis: CAD, finite elements, NURBS, exact geometry and mesh refinement[J]. Computer Methods in Applied Mechanics & Engineering, 2005, 194(39/40/41): 4135-4195.

第 9 章 论文投稿

9.1 期刊的选择

1. 重要性

英文科技论文写作的目的是学术上的交流。在研究成果完成后,在合适的国际期刊上发表论文可以方便与同行充分交流学术思想和研究成果,同时也使作者能够实时掌握本学术领域的研究动态,并不断提升作者的研究水平,促进科学技术的发展,从而更好地为人类社会服务。因而,选择一个合适的国际期刊来发表作者的研究成果至关重要,期刊的选择从一定程度上反映了受众群体和论文水平[1-3]。

SCI 是美国科学情报研究所创办的世界著名期刊文献检索数据库,能被 SCI 检索到的期刊就是 SCI 刊源。SCI 期刊的质量从一定程度上反映出了它所录用论文的质量,发表后读者的数量以及论文的引用率等。一般说来,质量越好的 SCI 期刊,其受众群体越大,期刊中的论文越容易被引用。

论文的被引用率是评价一个学者学术水平的重要指标之一。因此,在高水平期刊上发表论文可以推动本学科的发展、有利于基金和项目的申请,有利于评定单位或国家的学术水平,有利于提高作者在本领域的学术知名度。此外,在优秀学术期刊上发表论文,从学生角度看,在申请国外的博士、博士后或者就业等方面会获得更多的机会。当然,优秀期刊对论文的创新性和写作水平要求很高。因此,审稿人对稿件的质量要求也非常苛刻,录用率较低。

SCI 期刊也明确了论文的投稿周期。除了各行业协会(比如 ASME、ASCE 等)创办的期刊外,目前 SCI 期刊的版权主要被三大出版商(Elsevier、Springer 和 Wiley)收购。其中 Elsevier 的期刊最多,目前工作效率最高,其审稿周期相对较短,一些非常优秀的论文甚至一两个月就能发表。因此,这几年 Elsevier 期刊的影响因子普遍都上升得很快。当然,Elsevier 期刊的质量参差不齐,期刊论文的审稿周期也长短不一,具体的审稿周期,可以通过登录期刊的主页,统计其发表论文的平均审稿周期作为参考。Springer 主要出版专著、会议论文集及部分期刊,其审稿周期处于 Elsevier 和 Wiley 之间,比 Elsevier 的期刊略长。Wiley 期刊的质量一般比较高,其审稿周期也较长,一审时间大部分在半年以上。其审稿周期长且论文录用后一般一个多月后网络出版,半年左右才见刊,Wiley 期刊的影响因子最近几年几乎没有太大变化。当然,行业内的学者对 Wiley 期刊的认可度比较高,Wiley 期刊邀请的审稿人都是业界知名学者,对稿件质量要求很高。

期刊的选择在很大程度上决定了论文被录用的概率。比如一般水平的论文投稿到较高质量的期刊上,则很有可能直接被拒稿,没有修改的机会,浪费大量时间,长期如此甚至会影响科研者的信心。所以,选择质量要求与论文研究水平相当的期刊来发表研究成果非常重要,这样既可以尽快将作者的科研成果展示给读者,也提高了科研成果转化应用的概率。

2. 方法

SCI 期刊的影响因子(impact factor)[①]**和行业内的影响力是选择投稿期刊的主要方法。**影响因子是反映期刊影响力和阅读量的一个重要指标,*Nature* 和 *Science* 的影响因子都在 30 以上,许多 SCI 期刊的主页上还会列出期刊的近五年影响因子。利用 *Journal Citation Reports*(JCR,《期刊引用报告》)检索该期刊的总被引频次和影响因子可以了解期刊的学术影响力。期刊的总被引频次和影响因子越高,则表明该期刊被读者阅读和引用的可能性越大,进而可推断该期刊潜在的学术影响力也越大[4,5]。

不同学科或专业领域的研究周期差别往往比较大,单一地采用影响因子去评价期刊的影响力并非完全准确的。有些期刊的影响因子可能比相近专业领域的期刊影响因子较小,但是其行业影响力非常大,甚至是行业内的顶级期刊。因此,可以根据期刊的行业影响力来区分期刊的质量。在此简要列举土木类、工程地质类专业领域的重要期刊。

(1) 计算力学领域公认的重要期刊有 *Computer Methods in Applied Mechanics and Engineering*、*International Journal for Numerical Methods in Engineering*、*Journal of Computational Physics*。

(2) 岩石力学领域公认的重要期刊有 *International Journal of Rock Mechanics and Mining Sciences*、*Rock Mechanics and Rock Engineering*。

(3) 土力学领域公认的重要期刊有 *Geotechnique*、*International Journal for Numerical and Analytical Methods in Geomechanics*、*Computers and Geotechnics*、*Journal of Geotechnical and Geoenvironmental Engineering*。

(4) 结构工程领域公认的重要期刊有 *Composite Structures*、*Computers & Structures*、*Structural Safety*。

(5) 桥隧工程领域公认的重要期刊有 *Tunnelling and Underground Space Technology*、*Journal of Bridge Engineering*。

(6) 工程地质领域公认的重要期刊有 *Geology*、*Engineering Geology*。

在选择期刊时,要根据论文的创新性以及发表的紧迫性合理地选择拟投期刊。论文的水平应该与期刊的质量相适应。如果论文研究水平高于拟投期刊的质量要求水平,则有可能造成新的研究成果却少有学者下载阅读的窘况;如果论文研究水平低于拟投期刊的质量要求,则很有可能被拒稿,甚至可能得不到较好的反馈意见。读者可以根据平时阅读文献的期刊来源以及稿件所引用论文的发表期刊来选择拟投期刊。如果稿件中所引用的论文没有

---

① 某个期刊在某年的影响因子的计算方法:该年引用该期刊前两年发表论文的总次数与前两年该期刊所发表的论文总数之比。

一篇是在拟投期刊上发表的,建议读者不要选择该期刊,因为稿件主题与拟投期刊要求的主题范围极有可能不相符合。此外,建议读者在投稿时多引用拟投期刊中与稿件主题相符合的论文。

以下是投稿前选择期刊时需要重视的问题或一些建议:
(1)投稿前要明确论文的受众对象,谁对你的研究工作感兴趣。
(2)投稿期刊的主题范围是什么,与本稿件主题是否相符。
(3)稿件的研究工作是理论型还是实践型,是否符合期刊的侧重点。
(4)拟选期刊的声誉和影响因子。期刊声誉是期刊总体质量的综合反映,包括:①编委会成员的学术地位;②编辑的业务素质;③期刊在学术界的影响;④读者的青睐度;⑤总被引频次和影响因子。
(5)稿件的研究水平是否与期刊质量要求水平相当。
(6)是否引用了拟投期刊的文章作为参考文献。
(7)拟投期刊中已经出版的论文质量和水平如何。
(8)其他因素,如审稿周期、论文出版速度、论文版面费、服务质量等。

9.2 投稿要求与投稿过程

在论文投稿前,首先要确定稿件是否在期刊的接收范围内,重视期刊的侧重点(比如有些期刊录用的论文以重大问题的理论分析和解析解为主,有的以新型数值方法为主,有的以重大工程实录或试验为主,有的以交叉学科或计算机辅助设计在其他领域的应用为主),避免稿件因主题不符而在初审后直接被拒稿。务必仔细阅读期刊的投稿要求,包括稿件的格式要求,比如投稿内容、篇幅、字号、字体、行间距、行号、公式、图表、参考文献写法等。一般说来,如果采用 Word 文档书写论文,表格和图片需要放在参考文献之后,紧接着放上文中提及所有图表,一面放一个。如果采用 LaTeX 撰写论文,则只需要按照期刊给出的模板撰写就可以,表格和图片仍在正文中间。如果格式细节没有处理到位,一般在稿件初审后编辑会将稿件退回要求作者重新修改。[6,7]

稿件的投稿过程大致分为以下几个步骤:
(1)选择投稿类型(article type)。我们通常写得最多的论文类型是 research paper 或 original paper,其他常见的文章类型有 technical note、review、letter、report 等。
(2)输入论文题目(title)、摘要(abstract)、关键词(keywords)、作者名以及作者的基本信息和联系方式。
(3)推荐的审稿人和需要回避的审稿人。期刊允许作者推荐认识的相关领域知名学者进行审稿,一般要求推荐两三位审稿人供主编或编辑备选,网络投稿过程中只需提供被推荐审稿人的研究领域和电子邮件即可。当然,为了避免过多熟人审稿可能会造成的不客观、不公正的审稿意见,期刊的主编或编辑一般仅采用一个作者推荐的审稿人,甚至一个都不采用。填写需要回避的审稿人应该具有非常中肯的理由,而并非稿件创新性或研究内容上的某种回避[8-10]。

(4) 需要给期刊编辑部或主编说明的其他问题。

(5) 上传 cover letter、highlights、manuscript、figures 和 tables 等文件。

cover letter 也称为投稿信，目的是告诉主编或者责任编辑本稿件的基本信息情况并做出该稿件的原创性声明，再列出通讯作者的联系方式等。典型的 cover letter 如下：

Laboratory for ＊＊＊

Department of ＊＊＊

Address

Email：＊＊＊

Date

Dear Editor,

I hereby submit a manuscript entitled "＊＊＊" to *Journal name*.

This is an original manuscript which has neither previously, nor simultaneously, in whole or in part, been submitted anywhere else. If possible, we hope it may be published in your journal. Its publication is approved by all authors and that, if accepted, it will not be published elsewhere in the same form, in English or in any other language.

Correspondence concerning this paper may be sent to me at the address above.

I am looking forward to hearing from you soon. Thanks!

Sincerely yours,

Name of the corresponding author

highlights 是对稿件创新点的介绍，一般写三四点就可以，一句话概括一个创新点。主要是方便主编或者责任编辑确定论文的创新性是否达到期刊的要求，确定研究范围和寻找合适的审稿人等。有些时候，highlights 也出现在发表的论文中，通过阅读 highlights，读者就可以清晰确定论文的创新点。

(6) 生成 PDF 文件并下载检查、修订。

(7) 提交并确认投稿，期刊系统收稿回执。经审稿修改后提交的审改稿在上传文件时还须包含对审稿意见的具体答复(point to point responses to reviewers)。

投稿状态可以在投稿系统中查阅，一般包含以下几种。

(1) submitted to journal。论文刚提交的状态，一般持续时间为一天至数天。

(2) with editor。如果在投稿的时候没有要求选择编辑,就先到第一主编手上,第一主编会分派给别的主编或者责任编辑,一般持续时间为一周左右,直到责任编辑找到合适的审稿人为止。

(3) under review。论文处于评阅人审稿状态,不同的期刊一般持续数周到数月不等。

(4) required reviews completed。期刊要求的所需审稿意见已返回。如果审稿意见全部返回,那么很快就会进入稿件的下一个处理状态;如果只是期刊要求的最少审稿意见已返回,那么,这个状态可能会持续较久。

(5) decision in process。责任编辑或者主编正在审阅评阅人的审稿意见并给出审稿决定。正常持续一天到数天,当需要所有责任编辑或者主编共同讨论决定的时候则会持续较长时间。

(6) minor revision/major revision。论文返回作者进行修改阶段,包括小修和大修两种。大部分期刊给出大约 45d 的修改时间,可以根据稿件修改的需要,向责任编辑写信申请延长稿件的修改时间。

(7) accepted/rejected。论文最终决定——接收/拒稿。

不管是修改还是接收或者拒稿,通讯作者都会收到主编关于审稿决定的邮件,其中也会附上评审人的审稿意见。审稿意见对作者提高论文的质量很有帮助。因此,不管意见的好坏都应该认真对待。有的时候在 decision in process 状态之后又会重新回到 under review 状态,这种情况往往是由于审稿人对论文的意见相左,责任编辑觉得文章有意义,因此,责任编辑需要通过增加审稿人,获得更多评阅意见,从而确定文章是否录用或者修改。

9.3 实例介绍

本书以 Elsevier 旗下的期刊 *Computer Methods in Applied Mechanics and Engineering* 的投稿过程为例,进行详细的投稿步骤介绍。

(1) 登录投稿系统。其界面如图 9-1 所示。点击期刊主页上的"submit paper"即可进入。输入用户名和密码后,点击"Author Login"。如果没有用户名,可以点击"Register Now"申请一个新的账户。常规采用通讯作者的账户进行投稿。

(2) 进入投稿系统主页。其界面如图 9-2 所示。"New Submissions"用于新论文的投递。"Revisions"用于修改完成后论文的投递。"Completed"列出已做出最终决定的稿件列表。

(3) 选择文章类型。其界面如图 9-3 所示。根据撰稿论文的类型可以选择"Research Paper""Short Communication""Review Article"等。

(4) 输入论文的标题。其界面如图 9-4 所示。

(5) 录入作者的信息。其界面如图 9-5 所示。其中作者姓名、邮箱地址和国别为必填项。

(6) 输入摘要,其界面如图 9-6 所示。可以直接从稿件中拷贝、粘贴。

图 9-1 投稿系统登录界面

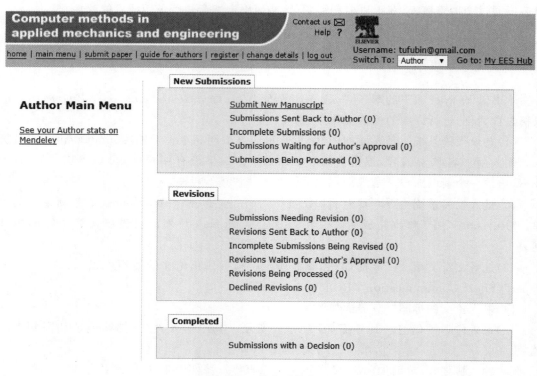

图 9-2 投稿系统主页

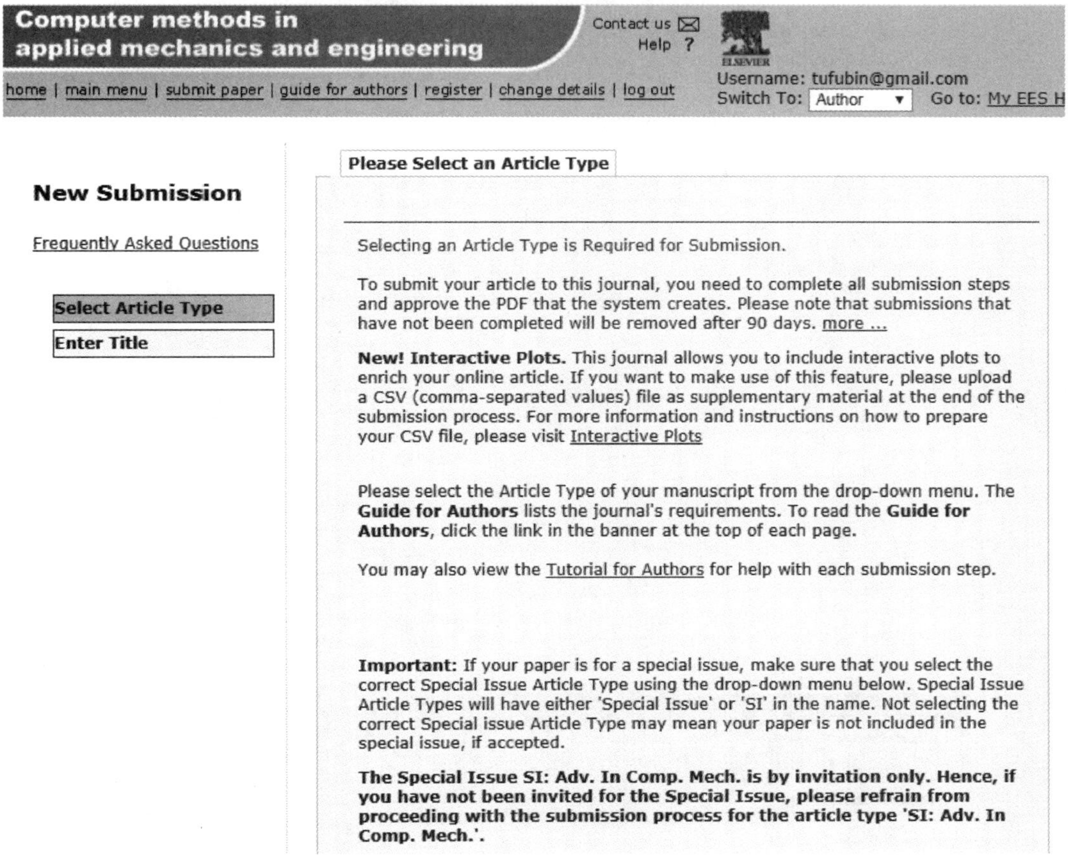

图 9-3 选取文章类型界面(a)和具体选项(b)

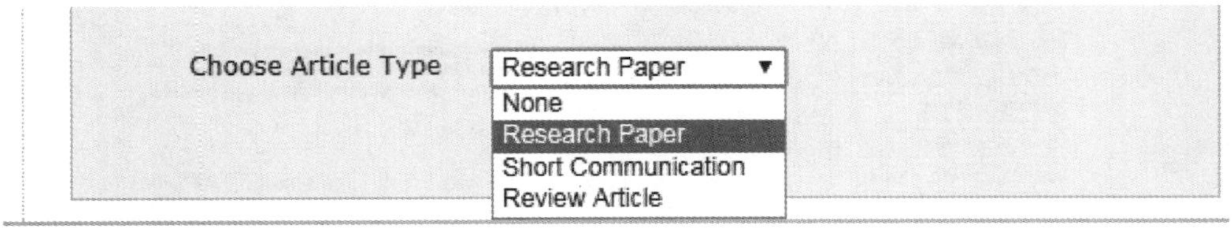

(7) 输入关键词。其界面如图 9-7 所示。一般要求不多于 6 个，关键词之间以分号相隔。

(8) 附加信息的确认。其界面如图 9-8 所示。需要确认的信息包括致谢中资助项目的完整信息、是否为方法论文且上传"MethodX"文件夹。

(9) 给出推荐的审稿人。其界面如图 9-9 所示。提供姓名、电子邮箱和推荐理由等。

(10) 给出宜予以回避的审稿人。其界面如图 9-10 所示。提供的信息同(9)。

(11) 选择主编或者责任编辑。其界面如图 9-11 所示。

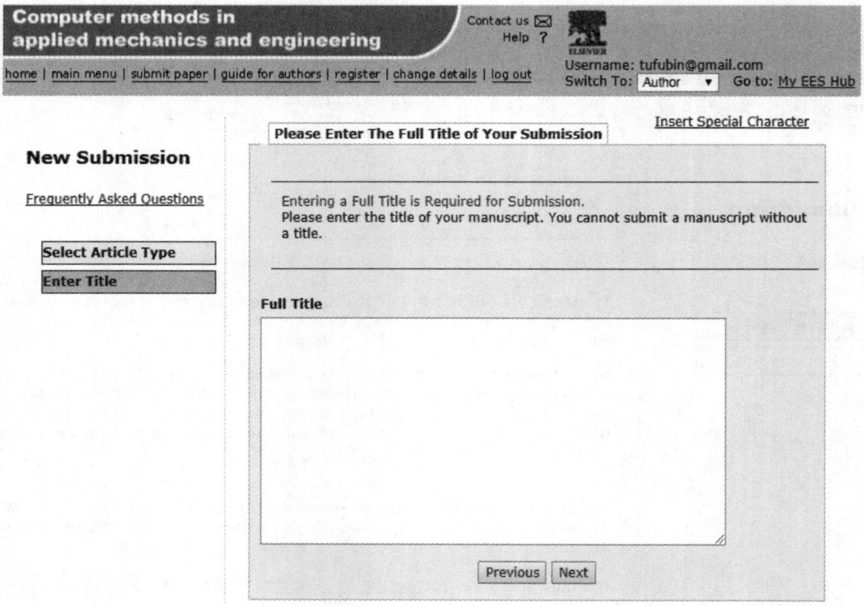

图 9－4　输入论文标题界面

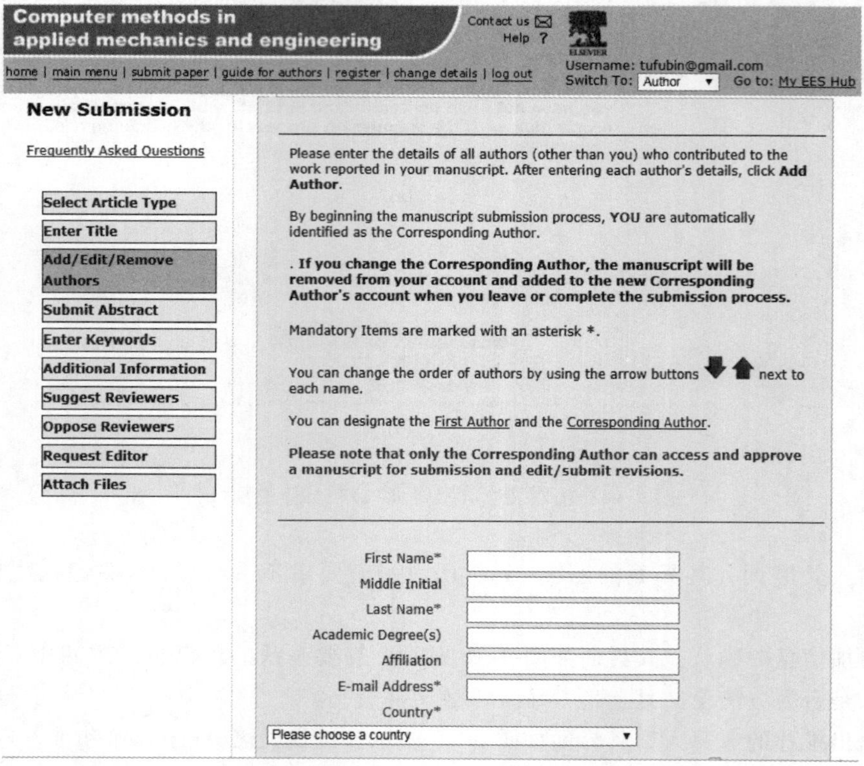

图 9－5　录入作者信息界面

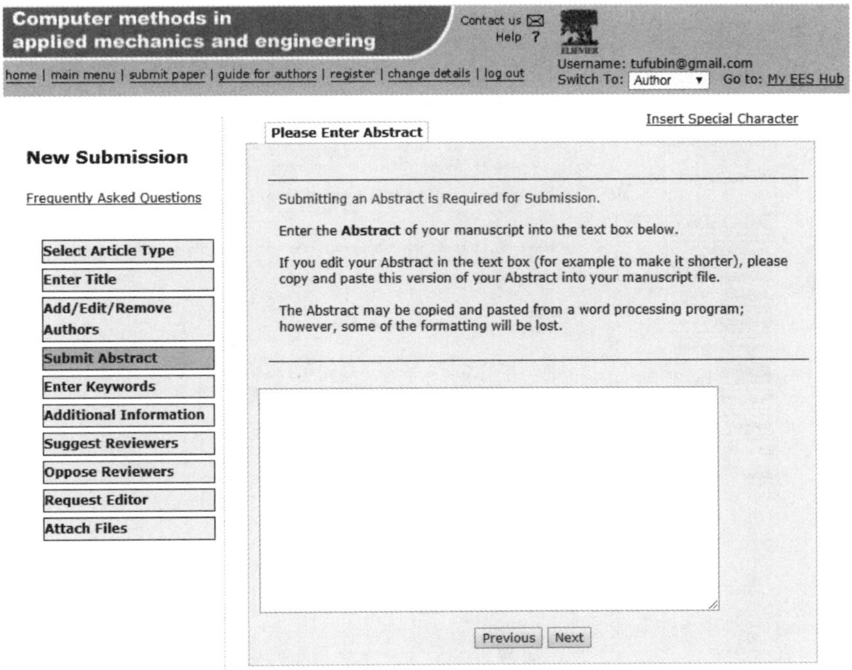

图 9-6 输入摘要界面

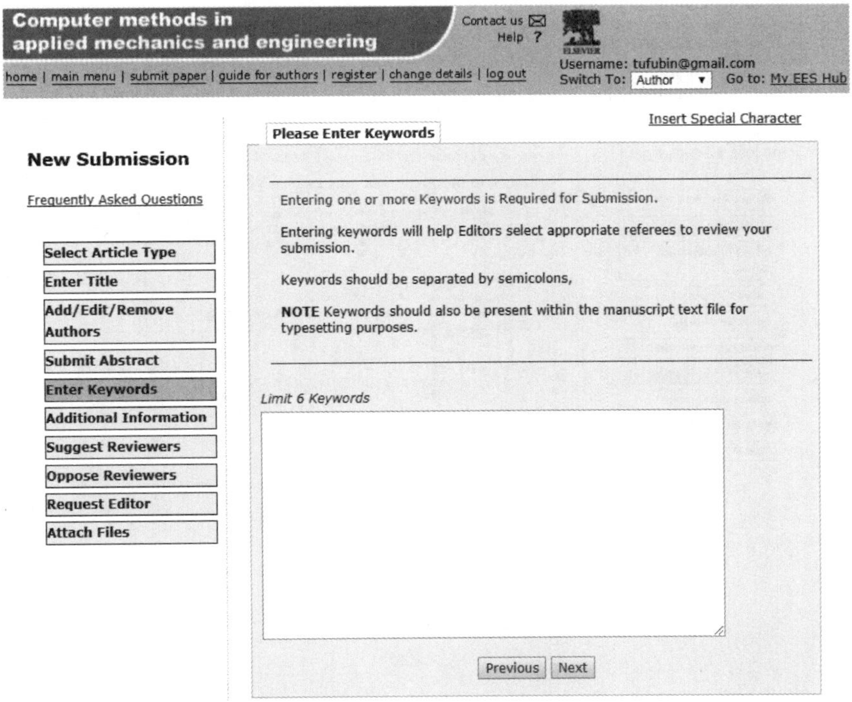

图 9-7 输入关键词界面

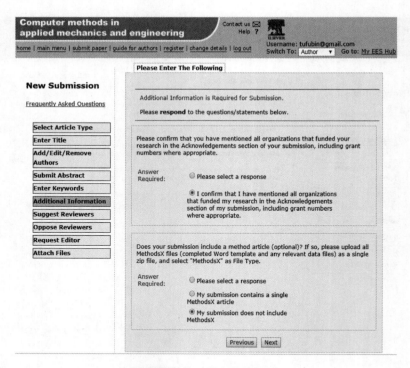

图 9-8　输入附加信息界面

图 9-9　推荐审稿人界面

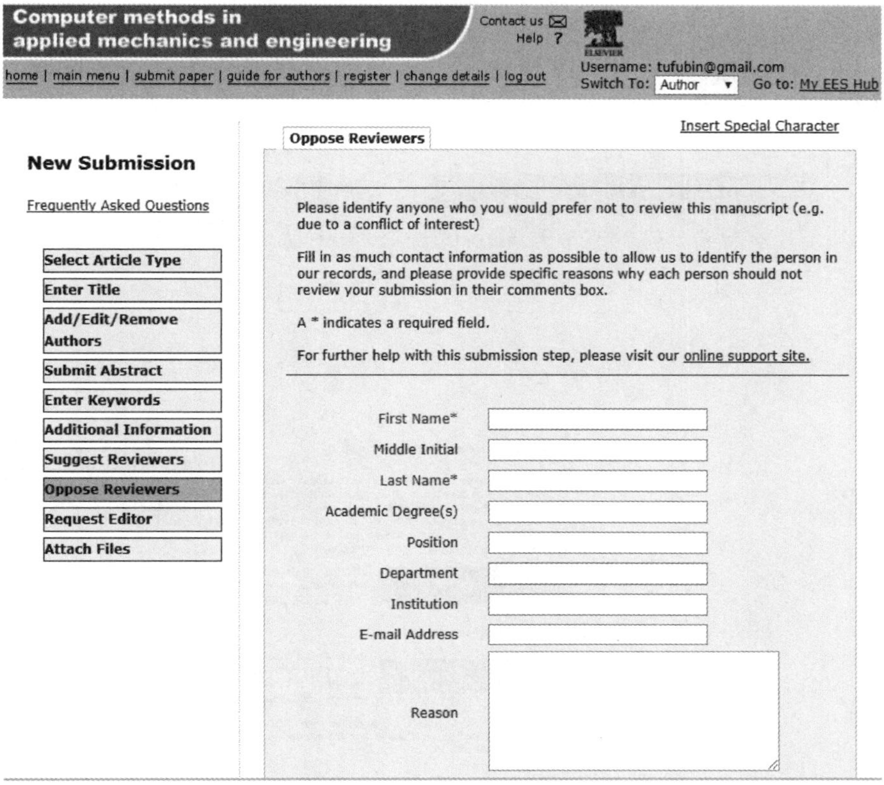

图 9-10　回避审稿人界面

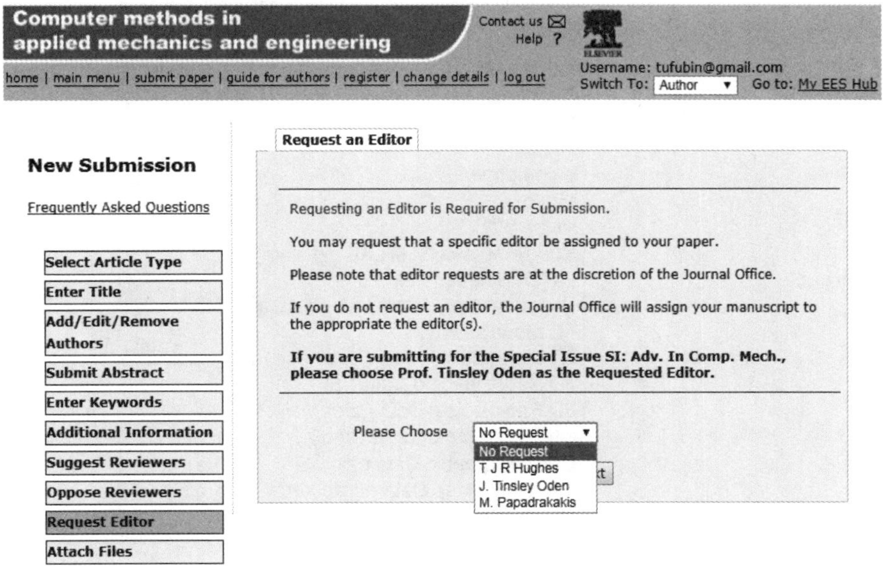

图 9-11　选择主编或者责任编辑界面

(12) 上传文件。其界面如图 9-12 所示。其中"Highlights"和"Manuscript"为必须上传的文件。它们的格式要求可以查阅各个 SCI 期刊的投稿须知。一般情况下，投稿信（cover letter）也是需要提供的。

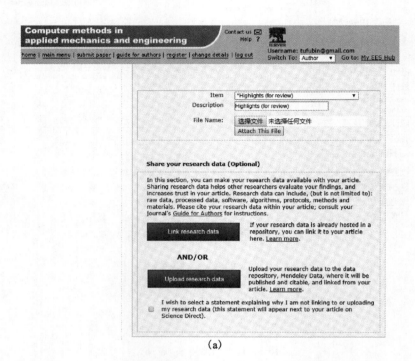

(a)

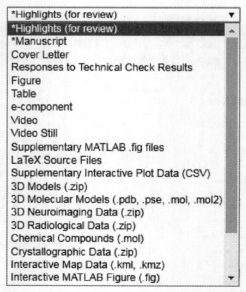

(b)

图 9-12　上传文件界面(a)和上传文件列表(b)

(13)上传文件的确认。其界面如图 9-13 所示。最后就可以生成 PDF 文件并确认无误提交。

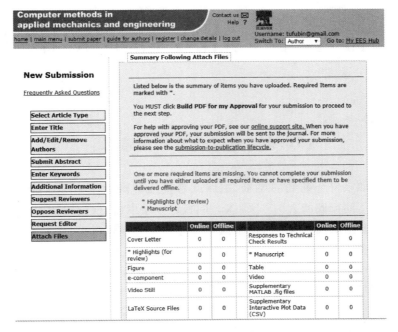

图 9-13　上传文件确认界面

本章参考文献

[1] 任胜利. 英语科技论文撰写与投稿[M]. 北京:科学出版社,2004.

[2] 张孙伟,吕伯昇,张迅. 科技论文写作入门[M]. 3版. 北京:化工工业出版社,2007.

[3] 孙玮,赵卫国,张迅. 科技论文写作入门[M]. 5版. 北京:化学工业出版社,2017.

[4] 欧阳晓黎,牛燕平,刘宏波. 从期刊评价指标看,如何提高向 SCI 源期刊投稿的命中率[C]. 中国科学技术期刊编辑学会. 第二届全国核心期刊与期刊国际化、网络化研讨会论文集. 北京:中国科学技术期刊编辑学会,2004:87-88.

[5] 林昌东. 国际著名期刊主编 Ferguson 教授谈如何向国际学术期刊投稿[J]. 中国科技期刊研究,2003,14(4):457-458.

[6] 李京华,张凤英. 如何向国外专业期刊投稿[J]. 中国科技期刊研究,2001,12(3):239-240.

[7] 姚逸. SCI 简介以及投稿应注意的问题[J]. 现代情报,2002,22(11):113-113.

[8] 昌兰,彭敏宁,陈丽文,等. SCI 及向其源期刊投稿的注意事项[J]. 中南大学学报(医学版),2005,30(3):370-372.

[9] 张润芝. 向 SCI 来源期刊投稿的技巧与方法[J]. 情报探索,2007(10):118-119.

[10] 金伟. 我国科技期刊投稿指南中存在的问题[J]. 中国科技期刊研究,2013,24(1):118-120.

第10章 论文评阅意见的反馈与修改

10.1 如何分析评阅意见

论文投稿后,如果符合期刊的要求,主编或者责任编辑会将论文送给评审专家进行评阅。在一般情况下,同行评阅人为3~5人。同行评审结束后,根据同行评阅人的评审意见,主编会给出以下几种审稿决定:直接录用(accepted),小修(minor revision),大修(major revision),拒稿后修改重投(rejected and resubmitted)以及拒稿(rejected)。一般来说,直接接收录用的论文非常少见,大部分的英文论文都要经过一定的修改才能达到期刊发表的要求。论文的撰写和发表就像雕琢一件工艺品一样,只有不断地进行精雕细琢,才能铸就精品,好的论文都是经过好几轮磋商,不断完善、修改出来的。而审稿意见就好比客户的要求一样,是使论文变得卓越的推动力。审稿意见越苛刻,改正后的论文就越完美[1,2]。

对于英文论文来说,只要给修改(minor revision 或 major revision),就有发表的机会。无论是小修还是大修,都要认真对待。在一般情况下,小修的文章修改后,主编认为修改到位的话,不会再返给审稿人评审,而是直接录用。而对于大修的文章,文章修改后一般都会返给审稿人重新评审,直到审稿人没有新的疑问为止。

不管是 minor revision 还是 major revision,一般只要出现以下字眼:"… can be published in the *Journal name* if authors can address the issues …""The paper may be acceptable for the publication after the required revisions are applied.""… there are some points that I ask the authors to address before the manuscript is accepted for publication." "After the following issues being addressed, the paper may possibly be suitable for publication in *Journal name*.""I hope the authors can revise(or possibly clarify)before it is again considered for publication."等都是非常正面的评审意见。只要作者按照审稿意见认真修改,肯定能被期刊接收。

对于大修的审稿意见,当审稿人质疑论文的创新性时,比如出现如下字眼:"But it is not enough as a journal paper for this publication.""I do not think the manuscript is suitable for publication in *Journal name*, and I do not think that the shortcomings can be fixed with a simple revision and re-review.""The presentation and demonstration of the method do not meet the standards of a journal like *Journal name*.""I do not recommend this paper for publication in *Journal* name.""On the whole, the paper suffers from several weaknesses …"等时,一般主编给的最终审稿决定是拒稿或者大修。如果是大修,则还有发表的机会,作者

一定要认真修改,尽量让审稿人满意,这样文章才有接收的机会。

一般来说,评审人的具体审稿意见可分为两类:疑问、要求(建议)。疑问型的评审意见可为疑问句的形式,包括以 be 动词或者助动词引导的一般疑问句和以"why""how"和"what"等疑问词引导的特殊疑问句。审稿人对数据、方法、理论推导和创新点的质疑等均属于疑问型审稿意见。有些疑问型审稿意见涉及论文的创新性,非常难以回答。比如,质疑稿件的研究"起点"与目标刊物的学术水平不相符;稿件研究内容的普遍性较差;稿件缺乏创新性或没有实用价值。要求或者建议型的评审意见,一般带有以下词汇,如 please、suggest、recommend、require、explain 等,可能涉及的具体问题:相关文献评述不够充分,缺少部分重要文献;所提出理论的支持数据不足,论证不充分;缺少对相关联的重要问题的讨论(discussion)或对应用局限性的讨论;缺乏具有概括性的结论(conclusion);语法或者表述错误;某些观点需要进一步的解释;图表、公式的修改;句子或者段落的删减等。相比于疑问型审稿意见,要求或者建议型审稿意见往往比较容易修改。关于上述意见,作者要做的主要是按照审稿人的意见认真修改,客观评述研究现状和存在的问题,补充数据,进一步分析原因,讨论结果,提炼结论。

评审人的审稿意见有轻重之分。比如,评审人针对文章创新性、文章的核心研究内容和结论所提出的意见常常就是最重要的,一定要给予高度重视;对于尚未直接拒稿的稿件,一般可以通过增加相应的研究工作来进一步提升稿件的研究水平;针对句子的表达方式、图表的格式等所提出的意见则可以认为是次要的意见,只需要做相应的修改就可以。

10.2 如何针对意见进行论文修改

对审稿专家提出的疑问和建议,一定要仔细分析、认真思考,确认哪些问题是必须进一步修改、完善的,哪些问题是需要讨论和说明的。前者往往对提高论文的水平和实用价值具有建设性的作用,后者对论文研究结果的适用范围和论文规范性具有重要作用。

论文的修改一般包含两个文件(revisions 和 responses to reviewers)。要求在 responses to reviewers 中对每一位审稿专家提出的每一个问题按照点对点回答(point to point response)的形式进行详细解答,并在修改稿中进行必要的补充和说明。论文中修改之处一定要采用特定的标注在修改稿中进行标注,并在 responses to reviewers 的文件中告诉审稿专家是如何做出相应修改的,具体的修改位置在文中的什么地方。

在论文修改过程中,要充分考虑审稿意见,一般尽量满足审稿人的要求。例如,审稿人推荐增加的参考文献,一般要引用,并进行客观评述,深入讨论。各种问题需要修改的程度可以根据审稿意见的用词判断,即 require>suggest=recommend>explain=clarify。对于审稿人提出的文中支持所提出理论的数据不足,论证不充分等问题,审稿人要求增加部分试验或者数值算例,在条件允许的情况下,一定要增加,并在修改稿中补充完善相关内容;若条件不允许,也要诚实地在 responses to reviewers 中进行说明,让审稿专家能够认识到要求增加的试验或者数值算例确实存在困难,难以实现,并表示在后期的研究工作中逐步推进。关于缺少对相关联的重要问题的讨论或对结果应用局限性的讨论等方面的问题,作者不应回避,应该结合文中的研究条件和结果对得出的数据和规律进一步分析讨论,讨论文中简化模型的必要性与合理性,阐明其结果的适用范围等;关于结论的提炼,作者需要注意的是研究结果并不等同于结论,结论不是数据结果的复述,而是对数据结果所展现的规律的提炼,结论具有高度概括性和普遍性。对于论文的表述错误,一定要仔细检查,认真对待,尽量避免。对于审稿人提出的语法、表达方式等问题,要按照要求进行修改。当审稿人提出改进语言问题的时候,可以寻找熟悉该领域的外国专家帮助修改,也可以找期刊推荐的润色机构进行润色。

值得注意的是,由于审稿人研究方向的差异或认识事物发展规律的角度不同,审稿人提出的所有问题并不都是完全正确的、合理的。对少数无法修改的意见或学术观点上的争议,作者要想办法提供充足的证据,比如利用已经得到的实验数据规律,或引用已有的重要文献,或者增加必要的试验或数值算例进行客观、合理地解释或反驳其学术观点,通过事实说服审稿人。即使无法说服审稿人,至少要做到文中的结果或结论具有充分的事实依据,让主编或者责任编辑能够感受到作者的研究工作是在限定条件下客观规律的真实反映。因此,主编和责任编辑会在后续的审稿过程中给予积极的判断,珍视学术争议的价值[3-5]。

针对多次被拒的文章,当论文质量与投稿期刊水平相当时,一般说来都是创新点不够或者核心研究内容的完整性存在问题。在这种情况下,一定要针对重要的审稿意见,想办法增加研究工作,增加创新点,丰富研究内容。针对多次被拒的稿件,作者一定不要灰心。稿件只要有中肯的评阅意见,其质量就有进一步提升的空间,完全可以根据评阅人的重要评审意见,把被拒稿的稿件当作大修的论文来修改,然后重新投稿。

总而言之,在修改过程中态度决定一切。审稿人、主编和作者的最终目的都是尽可能地提高文章质量,使修改后的文章达到期刊发表的要求。

10.3 对评阅意见的答复

在对评阅意见进行答复时,首先要对审稿人和主编(或责任编辑)对提高论文质量所做出的贡献表示感谢,告诉审稿人和主编(或责任编辑)已经对所有的审稿意见进行了点对点的答复,并在论文修改稿中进行了详细的标记[6,7]。

responses to reviewers 的第一段可以写成"Thank you for investing the time to review our manuscript. Your comments helped us improve our paper and we believe that we have a better paper now. The following lists the changes made according to your suggestions and a few other minor modifications. Each of the issues raised was numbered and clearly replied to. *Text in italics are quotations from your reports.* When appropriate, we have identified the lines numbers in the revised manuscript where the changes were made. Modifications and additions were highlighted in yellow in the revised manuscript."

针对审稿专家提出的疑问进行答复的格式一般为：

Issue ♯ ...

Why ……？

Reply：……. We have added the detailed explanation in the revised manuscript, see Page ×, Lines ×$_1$-×$_2$.

针对审稿专家提出的要求或建议进行答复的格式一般为：

Issue ♯ ...

Stress contour should be added to ……

Reply：We totally agree with you, we have added stress contour. Please see Page ×, Figure ×.

对与审稿人不一致的观点，尽量采用柔和的语气指出审稿人观点的适用情况，再说明稿件并不适用于审稿人指出的情况，但文中的结果或结论一定是具有充分的事实依据。总之，在答复过程中，一定要尊重审稿人的审稿意见，做到客观、有礼貌，尽量避免与审稿人发生书面争执。当审稿人的意见出现错误的时候，尽量客观、委婉地以讨论的方式对审稿意见进行答复。任何情况下，都切忌使用不同审稿人相左的意见来答复另外一位审稿人的质疑。

10.4 如何与评阅人及编辑进行交流

在投稿过程中，作者不可避免地要与主编、责任编辑或者论文审稿人进行交流。作者与主编或责任编辑的交流主要包括：催稿、让责任编辑宽限论文的修改时间、关于论文修改的一些说明等。与审稿人的交流主要是学术观点上的交流。由于评阅过程一般是单边盲审，作者不知道评阅人是谁。在通常情况下，作者与审稿人的交流主要是通过审稿人的 comments 和作者的 responses to reviewers 来进行沟通的。有时候，一篇稿件会有好几轮的修改过程，这些修改过程中的 comments 和 responses 就是作者与评阅人不断进行学术交流和学术讨论的过程[8]。

催稿信的一般格式为：

Dear ＊＊＊，

I am writing this letter to check the status of our manuscript entitled "＊＊＊"(Manuscript Number, MN：＊＊＊) submitted to *Journal name* ＊＊＊ months ago. From the online system, we found that the manuscript is still in "under review". Could you please improve the review process for us? If possible, can you help us contact the reviewers to feedback the comments?

Thank you very much for your warm work and help. We are looking forward to hearing from you soon.

Best Regards,

＊＊＊

宽限论文修改时间信件的格式与上述信件差不多。只需要把其中的内容修改为：

I am writing this letter associated with our manuscript entitled "＊＊＊"(Manuscript Number,MN:＊＊＊) submitted to *Journal name*. We are trying our best to revise it. However,we need more time to make it perfect. Can you give us ＊＊＊ days and extent the resubmission deadline? Thank you very much!

关于论文修改的一些说明可以在投递论文修改稿的 cover letter 中进行说明,其格式与 cover letter 相同。只需要把其中的内容修改为：

I hereby resubmit the newly revised manuscript entitled "＊＊＊" to the *Journal name*.

We would like to thank for your kind letter on ＊＊＊ and the referees' report. We have carefully taken the reviewers' comments into consideration and amended the relevant part in the manuscript. Please find here enclose the newly revised manuscript and a detailed documentation in response to the referees.

I am looking forward to hearing from you soon. Thanks.

10.5 实例介绍

案例一：

这是一篇发表在期刊 Computers and Geotechnics 的论文 *The Impact of Soil Properties on the Structural Integrity of High-fill Reinforced Concrete Culverts*[9] 的投稿和修改经历。该论文两次投稿 Computers and Geotechnics 期刊都被拒,第三次继续投该期刊,之后经过两次修改的反复过程,最后被录用发表。以此为例阐述如何分析和处理论文的评阅意见。

稿件 *The Impact of Soil Properties on the Structural Integrity of High-fill Reinforced Concrete Culverts* 首次投递后,当天收到主编的意见"Completed-Reject",由于语言表达方面的问题而拒稿,主要意见如下：

Thank you for submitting the above paper to Computers and Geotechnics *for review.* Computers and Geotechnics *is a well-established international Journal that receives a large number of papers for possible publication. Papers of only the highest technical standard, written in clear and concise English, are published. I regret to advise you that the above paper needs extensive professional language editing to improve the quality of the English before it can be sent to reviewers for comment.*

经过三周的语言编辑润色和内容上细节完善之后,该稿件再次投递到该期刊。又经过三周左右的审稿周期,该稿件再次被拒稿。主编意见如下:

Your paper has been reviewed (see comments below). In light of the reviewers' remarks and rankings, I regret to advise you that the paper is unsuitable for publication in Computers and Geotechnics.

审稿专家主要意见如下(分析重要意见):

Reviewer #1: Comments on the manuscript COGE-D-12-00280

The paper deals with several aspects of the structural damage that can affect culverts.

The introduction very briefly presents the results of a survey on a set of 102 highway culverts, together with a review of the literature. Section 2 deals with the bearing capacity of the ground under the culverts, except for Subsection 2.5, in which experimental results from an instrumented culvert are presented. Section 3 discusses the possible beneficial or prejudicial effects of reinforcing the ground below the culvert foundation, on the basis of numerical simulations. Section 4 presents measured values of settlements for the same culvert as in Section 2.5. Section 5 states very clearly the conclusions drawn by the authors.

On the whole, the paper suffers from several weaknesses:

(1) A very important weakness is the lack of precision in the description of the numerical simulations (applied loads, sequence of simulations steps).

(2) Section 2.4 is interesting but very poorly written.

(3) It is difficult to understand why the measures presented in Sections 2.5 and 4 are not compared with numerical simulations. Maybe the intended scope of the paper is too wide, and as a result the connection between these sections and the rest of the paper is loose. Also, it seems clear that the title should be rethought in order to better reflect the contents of the paper.

This is the reason why I think that the paper is not suitable for publication in its present state, in spite of its interest.

In a more detailed way, many points could be improved.

这位审稿人后面还列出了一些关于细化研究方法的描述、数值结果和试验结果的对比、对文中规律的进一步解释说明、计算参数的来源和补充、术语表达和图表格式方面的问题。

Comments #1 分析:第一位审稿人意见分析了稿件内容,提出了很多 *major issues* 和 *minor issues*。稿件中确实存在很多核心内容和细节上的问题,因此被拒。同时,comments 中还提到"*This is the reason why I think that the paper is not suitable for publication in its present state, in spite of its interest.*",这就从侧面反映出了稿件的研究价值,通过进一步的修改还有希望在同等质量的期刊上发表。

Reviewer #2: The paper presents some interesting results, but it suffers from several weaknesses：

(1) *Fig. 1: how can the load be applied on the culvert directly when there is soil above? It would be more meaningful if a more realistic design load from traffic is applied on the ground surface above the culvert.*

(2) *The reviewer does not think it is meaningful to investigate the cases of $d < h$ (Fig. 2). Without any backfill above, the culvert is no longer a culvert; it is just an exposed pipe.*

(3) *Related to the first two comments; if the load is on the ground surface, the backfill should have a better influence on bearing capacity due to the soil arching effect, unless failure occurs in the backfill above the culvert.*

(4) *The authors are advised to compare their results with predictions from bearing capacity theory. At least from the conclusion, it seems that bearing capacity theory will predict similar trends.*

(5) *How can one interpret the result related to soil consolidation and apply it in design? It is easily understood that the bearing capacity would increase if the excess pore pressures dissipate. The important thing is how to estimate the strength increase in practice.*

(6) *The effects of soil stiffness and non-uniform soil deformation on the stress in the culvert are also common sense. What the authors should focus on is some formulae to estimate such effects* for practical design.

(7) *Field instrumentation in section 2.5 should be used to calibrate the numerical model, which was not done. The calibration should also be at the beginning of the manuscript.*

Comments #2 分析：从第二位审稿人意见来看，*comments* 中提出了七个重要的问题，前三个问题是相关的，其中还有一个与研究内容不一致的观点，这就需要结合稿件研究内容的实际情况，进行详细的解释说明。在七个问题中，三次提到"*how*"，也就是要详细论证如何去实现它，修改过程中需要补充大量的资料及解释说明；两次提到"*apply*"问题，强调研究成果如何应用到实际工程设计中；另外还提及一个验证问题，明确指出应该利用试验结果验证数值结果，而且还应该放在文章的前面。这些问题都是属于 *major issues*，该文被拒符合该期刊对论文质量的要求，也在情理之中。同时，该 *comments* 中还提到"*The paper presents some interesting results.*"，这条意见坚定了作者继续修改、完善稿件的信心。

作者根据上述意见和建议，对该稿件的内容进行补充完善，经过三个月的修改之后，第三次投稿 Computers and Geotechnics。又经过两个月的评审，返回评阅意见如下：

主编总体意见为"*Major Revisions*"：

Reviewers have now commented on your paper (see below). In light of these comments, MAJOR REVISIONS are needed.

Please include a list of changes or a rebuttal against each point which is being raised when you submit the revised manuscript.

审稿专家的主要意见如下：

Reviewer #1: *Manuscript COGE - D - 12 - 00388*

The paper presents numerical simulations aimed at analyzing …, *the paper is concise and the authors state their conclusions clearly.*

(1) *The authors' statements regarding the influence of the backfilling rate are somewhat obscure, and* should be improved, *it seems that the conclusions are only valid for the short-term bearing capacity of the subgrade layer.*

(2) *Another possible improvement would be* to define more precisely the quantities under discussion.

(3) *The last problem is that the reader may wonder why the numerical simulations of Section 3 are carried out for a geometry that is different from that of the field test (Sections 1 and 2) and of the simulations of Section 4.*

The paper is not suitable for publication, but could be accepted provided that relatively minor modifications are made.

Here is a detailed list of points to be improved or clarified.

该审稿人后面还列出了论文各章节中存在的一些细节问题，具体包含增加参考文献，变量符号表示的意义，语言表述的准确性。

Reviewer #1 总体上对稿件的研究工作是一种肯定的态度，如果能够完善审稿人提出的问题，稿件是可以被录用的。该审稿人提出了三个主要问题：一是关于加载速率的影响需要深入分析，它反映的结果是路基承载力的一种短期效应，对此需要进一步解释说明；二是讨论中更精确地进行量化分析，不宜简要阐述影响规律；三是数值计算的模型与现场试验的几何尺度应该一致。这三个问题都能够很好地帮助作者提高论文质量和内容的完整性，从作者修改的角度看也是易于实现。

Reviewer #2: *Your work is very interesting. Please undertake revisions to address the following suggestions, questions and concerns*:

(1) *Table 1 Survey results of collapsed culverts needs* much more explanation.

(2) *your use of the term "ground mechanical properties" should be changed throughout*; perhaps you mean *"soil properties"*.

(3) *I prefer to* avoid qualitative statements *like "reasonable agreement with the field test data"; I much* prefer a quantitative comparison *where you give the % difference in measurement and calculation.*

(4) *You have used your computer analysis to examine various geometrical and material characteristics for the soils around the culvert, and I found that aspect of the study interesting. However, some of your choices were a bit puzzling. For example, you did not examine box culvert wall thickness, which has a large effect on slab stiffness, and I think those changes in slab stiffness can lead to substantial changes in loads.*

(5) *All your analysis for the RC slabs assumes elastic response, whereas RC elements are designed to crack and the flexural stiffness drops dramatically after that point; you*

need to discuss this clearly in the text and indicate whether your calculations will be conservative or nonconservative.

除了上述主要问题，Reviewer #2 还提出了在影响因素分析中进一步考虑结构尺寸及刚度的影响，补充插图说明结构尺寸和位置关系。

Reviewer #2 的意见首先肯定了对论文研究工作的兴趣，提出了五个主要问题：第一个问题是对文中描述某种现象发生程度的具体解释；第二个问题是关于自定义属于的准确表达；第三个问题与 Reviewer #1 的第二个问题相同，希望对文中的数据结果采用定量描述而不是定性分析；第四个问题是关于主要内容的补充，进一步分析刚度的影响规律；第五个问题涉及应用价值，讨论文中理论分析结果与实际应用之间的差异。

各审稿人的主要意见有相同之处，也有很多互补的意见和建议，对文章内容的充实和应用价值等方面具有很好的提升作用。经过两个月的修改和一个月的审稿，期刊的返回意见为"Minor Revision"，涉及一个术语表达和两个自定义变量含义的解释。修改之后，经两周左右的时间，反馈信息为"Accept"！

从该稿件的投稿经历和修改过程来看，高水平国际期刊的审稿意见非常中肯，无论投稿结果如何，对专家的审稿意见认真分析，仔细修改，依据研究结果合理解释现象，对论文质量的提高具有很大的帮助。只要功夫深，就能改出一篇高质量论文。

案例二：

这是一篇发表在 International Journal of Mechanical Sciences 的论文 Discrete element-periodic cell coupling model and investigations on shot stream expansion, Almen intensities and target materials[10] 的部分审稿意见。以此为例说明 Responses to reviewers 的书写和稿件修改的标注问题。该论文经过一次大修后就被录用发表了。

论文的审稿意见如下：

Ref: SUBMIT2IJMS_2017_2934

Title: Discrete element-periodic cell coupling model and investigations on shot stream expansion, Almen intensities and target materials

Journal: International Journal of Mechanical Sciences

Dear ***,

Thank you for submitting your manuscript to International Journal of Mechanical Sciences. I have completed the review of your manuscript and a summary is appended below. The reviewers recommend reconsideration of your paper following MAJOR REVISION. I invite you to resubmit your manuscript after addressing all reviewer comments.

When resubmitting your manuscript, please carefully consider all issues mentioned in the reviewers' comments, outline every change made point by point, and provide suitable rebuttals for any comments not addressed.

I look forward to receiving your revised manuscript as soon as possible.

Kind regards,

K. Davey
Editor-in-Chief
International Journal of Mechanical Sciences

Comments from the editors and reviewers:

- **Reviewer 1**

This paper under the Ref code(SUBMIT2IJMS_2017_2934)can be published in the IJMS if authors can address the issues have been raised as below:

……

(10)How did you choose the different amounts for alpha parameters? (Page 13,Line 240)

- **Reviewer 2**

The paper presents a coupled DEM - FEM methodology to investigate the effects of the shot peening process on residual stresses and roughness by considering conical shot stream. Authors comprehensively investigated the effects of the shot stream expansion on the residual stress profile, roughness and shot impacting velocity and angle for different target materials, shot types and Almen intensities. The paper may be acceptable for the publication after the required revisions are applied. The comments are:

(1)The last paragraph of the introduction section, Lines 78 to 83 of Page 4, should be removed from the manuscript.

(2)There are some clerical errors in the manuscript. Some of them can be seen in Line 14 of the abstract section(results), line 127(H is the), Line 185(Figure 3b)and Line 299 (than) of the manuscript. Please review all over your manuscript and make sure that all sentences are correct.

……

(6)As mentioned in the manuscript, the tensile residual stresses depicted in Figures 13 to 16 are near zero due to using a periodic cell in simulations. So how the compressive residual stresses balanced with tensile stresses and how the equilibrium of the stress distribution in the periodic cell is occurred?

- **Reviewer 3**

This paper provides a DEM_FEM model to simulate the process of shot peening. It is quite detailed and investigates the effects of multiple parameters that can affect the results regarding the distribution of residual stress. The paper is quite well written and is of interest to the readership. However, there are some points that I ask the authors to address before the manuscript is accepted for publication:

(1) In the literature review, few detailed simulations of shot peening are missing. I suggest adding those to complete the state of the art section: a numerical model of severe shot peening(SSP) to predict the generation of a nanostructured surface layer of material, Surface and Coatings technology, 204, 4081 – 4090, 2010. Elasto-plastic pseudo-dynamic numerical model for the design of shot peening process parameters, Materials and Design, 30, 3112 – 3120, 2009. Finite element modeling of shot peening process: prediction of the compressive residual stresses, the plastic deformations and the surface integrity, Materials Science and Engineering A, 25, 173 – 180, 2006.

......

(6) The use of periodic cell is mentioned to result in null tensile residual stresses. However, the experimental data are not showing zero tensile residual stresses. Thus I guess this could not be considered as an advantage of periodic cell models.

......

尽管主编给出了大修的审稿决定,但从三份审稿意见来看,审稿人对文章的评价非常正面,都建议修改后发表。显然,第一位审稿人的第十条、第二位审稿人的第六条、第三位审稿人的第六条意见为疑问型的审稿意见;第二位审稿人的第一条和第二条、第三位审稿人的第一条意见为要求或建议型的审稿意见。第二位审稿人的第六条、第三位审稿人的第六条意见是对研究内容和创新性的质疑,为重要的审稿意见,必须增加研究工作做出答复和修改,为此,该论文增加了一些数值算例;其余列出的审稿意见为次要的审稿意见,只需解释或者根据建议进行修改即可。

针对以上审稿意见,Responses to Reviewers 的相应部分如下:

Responses to Reviewers

Discrete element-periodic cell coupling model and investigations on shot stream expansion, Almen intensities and target materials

by

Fubin Tu, * * *

Dear reviewers,

Thank you for investing the time to review our manuscript. Your comments helped us improve our paper and we believe that we have a better paper now. The following lists the changes made according to your suggestions and a few other minor modifications. Each of the issues raised was numbered and clearly replied to. *Text in italics are quotations from your reports*. When appropriate, we have identified the lines numbers in the revised manuscript where the changes were made. Modifications and additions were highlighted in yellow in the revised manuscript.

Reply to Reviewer 1

......

Issue #10

How did you choose the different amounts for alpha parameters? (Page 13, Line 240)

Reply to Issue #10

We fixed the β value as 2.0×10^{-10} s and changed the α value to ensure that the elastic wave fully dissipated during the impacting time interval. We set $\alpha = 1.0 \times 10^6 \text{ s}^{-1}$ first and increased it by 2^k till the requirement was fully satisfied. We found that $\alpha = 4.0 \times 10^6 \text{ s}^{-1}$ was sufficient. We have added the explanation in Page 17, Lines 242–245.

......

Reply to Reviewer 2

Issue #1

The last paragraph of the introduction section, Lines 78 to 83 of Page 4, should be removed from the manuscript.

Reply to Issue #1

This paragraph has been removed.

Issue #2

There are some clerical errors in the manuscript. Some of them can be seen in Line 14 of Abstract Section (results), Line 127 (H is the), line 185 (Figure 3(b)) and Line 299 (than) of the manuscript. Please review all over your manuscript and make sure that all sentences are correct.

Reply to Issue #2

Sorry for this mistake. We have corrected the errors and carefully read the manuscript again to remove grammatical errors.

......

Issue #6

As mentioned in the manuscript, the tensile residual stresses depicted in Figures (13) to (16) are near zero due to using a periodic cell in simulations. So how the compressive residual stresses balanced with tensile stresses and how the equilibrium of the stress distribution in the periodic cell is occurred?

Reply to Issue #6

This is very relevant comment indeed that required us to significantly change our modeling strategy and incurred months of simulation.

In the cell with conventional periodic boundary conditions (PBC), the compressive residual stresses are not equilibrated since implementing Equation (6) yields a 0 average strain. This is the same situation for the confined cell as well since it is typically clamped on its boundaries. A modified periodic boundary condition was proposed by using plane section assumption for a classical beam to obtain balanced forces and moments. We have added the description of modified PBC, see Section 3. 4.

The model with modified PBC requires to use the real thickness of the target and much more time to perform a simulation. We have proved that the results from modified PBC model converges to that from conventional PBC model as the thickness of a metal part increases. For the samples used in this paper, conventional PBC model is sufficient. See the analysis details in Section 4. 4.

Reply to Reviewer 3

Issue #1

In the literature review, few detailed simulations of shot peening are missing. I suggest adding those to complete the state of the art section: A numerical model of severe shot peening (SSP) to predict the generation of a nanostructured surface layer of material, Surface and Coatings technology, 204, 4081 - 4090, 2010. Elasto - plastic pseudo-dynamic numerical model for the design of shot peening process parameters, Materials and Design, 30, 3112 - 3120, 2009. Finite element modeling of shot peening process: Prediction of the compressive residual stresses, the plastic deformations and the surface integrity, Materials Science and Engineering A, 25, 173 - 180, 2006.

Reply to Issue #1

According to your suggestion, we have added the recommended papers in the introduction. See Page 3, Lines 39, 49, 56 - 58.

……

Issue #6

The use of periodic cell is mentioned to result in null tensile residual stresses. However, the experimental data are not showing zero tensile residual stresses. Thus I guess this could not be considered as an advantage of periodic cell models.

Reply to Issue #6

We have addressed this issue in Reviewer #2, issue #6.

……

修改后的部分论文如下：

Meguid et al.[12,13], Majzoobi et al.[14], frija et al.[15] and Kim et al.[16] *relied on symmetrical cells where normal displacements on symmetry boundaries were fixed to 0. The shortcoming of symmetry cells is that shots are required to impinge away from the boundaries or at points where their midpoints are exactly on the corners or sides. Moreover, the shots must be symmetrically cut into several pieces if they hit near the cell boundaries. Some authors claim that a symmetrical cell is not suitable for modeling random impingements*[17]. *Finally, symmetry cells can only deal with normal impacts, which is not compatible with the objectives we set for this paper.*

Miao et al.[18], Bagherifard et al.[19], Gariepy et al.[20], Mylonas et al.[21], Mahmoudi et al.[22] and Tu et al.[2] adopted a confined cell with fixed boundaries or infinite elements to simulate the randomly distributed shots impacts. In these models, the shots randomly impinged in a confined area of the cell surrounded by a marginal region that was used to reduce the edge effects. However, it was found that the tensile residual stresses were affected by the width of the marginal region[2,20]. Such a strategy requires convergence studies in order to properly size this marginal region. Shivpuri et al.[23] used a quarter of confined cell to simulate a single shot impact and compared the result with experimental measurement.

......

3.4. *Modified periodic boundary condition*

Conventional PBC constrains the movement of $+x$ or $+z$ surface to be the same as that of $-x$ or $-z$ surface. This particular situation prevents the resulting forces and moments to be balanced. It yields near zero tensile stresses that can not balance the compressive residual stresses[17,24]. *A modified periodic boundary condition was proposed to address this short-coming. The formulation assumes that the $\pm x$ and $\pm z$ surfaces remain plane, as in the beam theory. Figure 4b schematizes the deformation mode that can be taken by the cube's sides. For the sake of illustration, consider the $\pm x$ surfaces. The $+x$ surface rotates a constant angle around z with respect to the $-x$ surface, where n is the total number of layers in the y direction.*

......

Similar treatment was done in the z direction. Note that the deformation mode is those of a classical beam and will lead to equilibrated forces and moments, which in turn will lead to equilibrated residual stress fields.

......

Figure 7 shows the displacement-time curve for a node located at the center of the top surface for a single impact. The shot was located on the corner of an IN718 cell peened at an Almen intensity of 8A with CW14 shot. *Figure 7a shows the response for undamped and damped simulations for different values of the mass-proportional factor α. $α = 1.0 \times 10^6 s^{-1}$ was set first and increased by 2^k while the stiffness-proportional factor β was constantly set to $2.0 \times 10^{-10} s$.*

......

4.4. *Influence of target thickness when using modified PBC*

The shortcoming of the conventional PBC is that the compressive and tensile residual stresses are unbalanced since near zero tensile stresses are obtained. Simulations we have performed show that residual stresses results are also sensitive to the target thickness. Figure 15 shows the comparison of residual stress profiles for parts peened with CW14 shots at an Almen intensity of 8A, for different thicknesses. The figure shows that the residual stress profiles converge for both implementations for an IN718 part whose thickness is greater than 10 mm. Table 8 lists the computational time for all of the investigated cases. Conventional PBC greatly relieves the computational burden. Therefore, the results from conventional PBC were deemed acceptable and used for the samples investigated in this paper, whose thicknesses are of 10.2 mm and 25.4 mm.

本章参考文献

[1] 金能韫,王敏. 英语科技论文写作与发表[M]. 上海:上海交通大学出版社,2020.

[2] GASTEL B,DAY R A. 科技论文写作与发表教程[M]. 8 版. 北京:电子工业出版社,2018.

[3] 郑福裕,徐威. 英文科技论文写作与编辑指南[M]. 北京:清华大学出版社,2008.

[4] 刘振海. 中英文科技论文写作教程[M]. 中国科学院研究生院教材. 北京:高等教育出版社,2007.

[5] 意得辑. 英文科技论文写作的 100 个常见错误[M]. 北京:清华大学出版社,2020.

[6] 张康. 英文科技论文写作与发表[M]. 北京:清华大学出版社,2020.

[7] 刘振海. 中英文科技论文写作教程[M]. 北京:高等教育出版社,2007.

[8] 张俊东,杨亲正,国防. SCI 论文写作和发表:You Can Do It[M]. 2 版. 北京:化学工业出版社,2016.

[9] CHEN B G,SUN L. The impact of soil properties on the structural integrity of high-fill reinforced concrete culverts[J]. Computers & Geotechnics,2013,52(7):46-53.

[10] TU F B,DELBERGUE D,KLOTZ T,et al. Discrete element-periodic cell coupling model and investigations on shot stream expansion, Almen intensities and target materials[J]. International Journal of Mechanical Sciences,2018,145:353-366.

第 11 章　科技论文学术交流

11.1　概　　述

目前,随着国家的经济实力越来越强,科研项目的资助力度越来越大,国内外学术访问和交流越来越多频繁。中国的学者越来越需要在国际专业学术平台上用英文来大力宣传自己的科技创新成果。由于中英文化、思维和生活方式的差别,中国学者用英文做科技学术交流的实际效果有时还不是很理想,难以让不懂中文的外国听众完全理解演讲的内容,进而他们会对我们的创新成果失去兴趣或缺乏认同感[1]。这样,中国学者的学术交流也就难以达到期望的效果。

因此,中国学者用英文演讲一定要让懂英文的外国人听明白并能理解演讲的内容和意思,要让他们感兴趣、认同我们的创新成果,进而在将来引用我们的创新成果[2]。

11.2　科技论文学术交流的重要性与形式

1. 重要性

科技论文学术交流的重要性主要体现在以下几个方面[3]。

(1)了解领域前沿。参加学术会议,短时间连续倾听若干学术报告,最直接的受益是快速了解本领域及相近领域的学术前沿,了解行业动态。掌握大家都在做什么,做到什么程度,有什么意义和价值。

(2)分享研究成果。参加学术交流会,将自己最新研究成果,给同行进行简要的汇报。从选题、制作PPT到报告和会后讨论是一个研究阶段工作的总结,可以让同行给自己提出建议,明确下一步做什么、怎么做。

(3)启发科研思路。在听报告的过程中,各种思想相互碰撞,能开拓我们的思路和视野,激发灵感,很多科研想法会灵光乍现,进而丰富、发展自己当前研究,优化自己学术体系。

(4)重新评估自己。在参加学术会议过程中,听别人报告、看别人成果也是重新认识自己、评估自己的过程。看到好的科研成果,会让自己羡慕,也会感到不足。同时,也会让自己看到自信,给自己力量。

(5)提高鉴赏能力。学术会议很短,大多两三天,但报告集中,是同行密集展示自己科研成果的时候。通过同行横向比较,可以知道哪些单位、哪些学者做的研究水平高,进而提高了科研鉴赏能力,提高了自己学术品味。在平常工作中,也可以对自己的工作有个较为系统的评价。

(6)结实学术朋友。学术会议是个平台,让相同、相似研究领域的人走到一起,彼此认识、彼此交流,成为日常生活中的朋友、学术上的挚友,增进学术存在感,同时,也为平时学术思路交流提供了分享与讨论的人。学而无友,孤陋寡闻。有三五个学术好友,科研道路会有趣得多。

(7)全面培养学生。让研究生参加学术会议,可以拓宽视野,增长见识,掌握研究动态,培养科研兴趣。领略领域内著名科学家的风采,给学生树立标杆,激励他们成长。参加学术会议通常会成为他们的美好回忆。

(8)短暂放松自己。参加会议期间,可以让自己生活节奏慢下来,看看会议所在地的风景,拍拍照片,品尝当地美食,可以三五好友聚在一起聊聊天、叙叙旧,还可以信马由缰地写一写自己喜欢写的文字。

2. 形式

总的来说,学术交流的形式主要有以下几类[4]。

(1)学术年会。学术年会是学术会议中一种制度性的会议形式,通常是定期(一年或多年)召开的一种大型综合性或主题型学术会议。学术年会一般具有主题性、高层次、学术性、参与性、开放性、规模性等特点。每届需要确定一个或多个会议主题;会议规模比较大,参加人员比较多,包含的学科范围和专业领域较广,所以需要进行较长时间的筹备,通常筹备周期都在一年以上;年会期间,除进行必要的大会报告外,还需要设置若干专题分会场或专题单元,以及大型展览、科普活动等,以供参会代表和公众选择性地参与。

(2)国际学术会议。国际学术会议是各个国家相关学术领域研究者聚集交流的学术交流形式。国际学术会议的主要与会者来自各个国家,这是识别国际学术会议的主要标志。换言之,国际学术会议的每个与会者都具有某个国家的象征,不论你是否是国家正式派遣的代表,还是以个人身份与会,人们都会把你列在你所属国家名下,把你看成是这个国家的一部分。由此,国际学术会议和国内学术会议的区别为:国际学术会议是不同国籍学者的会议;国内会议是同一国籍学者的会议。国际学术会议分双边学术会议和多边学术会议。双边学术会是指与会者仅来自两个国家,多边学术会议是指与会者来自三个及三个以上国家。

(3)学术报告会。学术报告会是指以介绍科技发展和学术研究动态、发布学术研究成果等为主要内容的学术演讲会。这种会议往往以演讲者为中心,听讲者参与讨论的机会不多。有时与其他学术活动相结合,具有科普和科技传播的作用。学术报告会的题目要确定,并事先告知,以便听众选择和有所准备;学术报告演讲题目要与听众的知识背景和学术兴趣相匹配。学术报告会一般采取影剧院式的会场形式,设报告台或演讲台,主持席可设可不设;应使用多媒体投影、音频视频系统,灯光调节要适度;单个学术报告时间不能过长,可以安排多位专家做学术报告;报告会的总时间比较灵活,可长可短。

(4)学术沙龙。沙龙是法语"salon"的音译,原指法国上层人物住宅中的豪华会客厅。17世纪,巴黎的名人常把客厅变成著名的社交场所,志趣相投者欢聚一堂,促膝长谈,无拘无束。后来,把这种形式的聚会叫作沙龙,学术沙龙由此衍生。典型的学术沙龙的特点:第一,会议定期举行;第二,会场自在宽松,会议氛围朦胧浪漫,常常在晚上举行,会场清静别致,以激起与会者的学术情趣、谈锋和学术灵感;第三,学术志趣相投,参会者人数不一定多,一般以20人以内为宜,但必须是沙龙学术主题的圈子内人,社会地位、年龄、性别等不限;第四,自由谈论和交流,设定一个交流主题或范围,可以集体谈论和交流,也可以自愿结合,三三两两,自由谈论,各抒己见。有的学术沙龙组织者将谈论、交流进行记录、整理,扩散和传播到更大范围。

(5)小组学术讨论会与专题讨论会。小组学术讨论会与专题讨论会是一个非常接近的学术会议形式,是指与会者为研讨某一专门性的学术问题或学术研究任务而进行的沟通、讨论、分享知识、技能和对问题的看法的聚会活动。小组学术讨论会一般是学术会议中基本的单元形式,而专题讨论会既可以作为某一学术会议的单元,也可以单独召开。小组学术讨论会分为两种形式,即有领导的小组学术讨论会和无领导的小组学术讨论会。

(6)学术论坛与学术讨论会。学术论坛是指学术团体或主办机构召集相关学术研究者就某一学术问题或论题,聚会到一起反复深入地研讨或论证,充分发表学术言论,并有听众参与其中的学术会议形式。学术论坛与学术讨论会、学术交流会等常常难以分开,多数时候只是名称的差异。学术论坛一般由主持人或演讲者自己主持,各方面对学术主题感兴趣的专业人员和听众均可参与,演讲者面对全体听众演讲,并就论坛主题发表意见和看法。当两个或更多的演讲者持不同学术观点、意见时,可以在演讲者之间、听众与演讲者之间展开自由、公开的讨论,允许听众提问。而学术讨论会、学术交流会有时不是面向全部听众,而通常只是几位与会者可以参加交流和提问。

(7)学术讲座。学术讲座(或学术讲授)与学术研修会是传授学术知识、获得学术知识的两个环节。学术讲座是一种知识、学术、资讯沟通的学术交流方式,通常采用学术主讲人当面演讲或网络、视频、音频、文本等演讲的方式把自己的学术知识分享给其他需要的学术研究者或学生、受众。现场学术讲座听众可以发问,互动讨论,面对面地双向沟通。学术讲座中,学术话题、主讲人、参与者是基本的构成要素,学术主讲人要最大化地让听众来分享自己的学术知识。

(8)学术会议墙报。学术会议墙报(poster)也称学术会议展板,是学术交流一种常见方式,与学术会议相配合,是学术会议的一道独特的学术风景。学术论文成果展板交流是一种在国际会议期间进行学术交流的常用形式,也是国外大学和研究机构开展学术研究成果交流与展示的一种普遍方式。

11.3 学术会议展板交流

会议展板展示已经成为很多学术会议的正式组成部分,有些会议甚至安排有多个展板展示时段。由于展板展示更有利于直接、快速地交流,方便作者与感兴趣的读者进行深入交流,并由此建立联系,因此得到很多会议参加者的认可[5]。

11.3.1　会议展板的结构布局

会议展板的布局要根据组织者的相关要求(展板的长度和宽度)及展演的内容和目的决定,要力求将自己拟传递给读者的信息以简洁明了、易于理解的形式表达出来。有些会议安排的展板有成百上千幅,因此,如何使展板具有吸引力,给人良好的第一印象是至关重要的。

通常情况下展板中的内容分为三四列;摘要在左上方,结论在右下方,参考文献、致谢等次要的文字安排在右下角;字体、字号和空白(行间距等)要适度变化,尽可能采用易于阅读的无衬线字体(如 Arial),题名文字要在 3m 以外能看清,正文文字应在 1m 以外能阅览(题名,90 号黑体字;章节标题,36 号字;正文,24 号字;1.5 倍行距),图表中的字号也应该足够大;尽可能多地使用图件或表格,文字、图表和空白的大致比例为 30%、40% 和 30%;展板中各模块的风格要尽量保持一致(标题的位置相对固定),信息的展示要有清楚的逻辑顺序[6]。

为丰富展板的表现形式,可适当使用背景颜色,但背景色一定要柔和,如果图件的颜色整体偏深,应使用浅色背景,否则使用深色背景;背景颜色要尽量一致,不宜过于鲜亮,不宜有图案,以免干扰读者;如果要突出重要内容的文字或图表,可适当使用与背景色对比明显的方框。

会议展板的版面设计通常有大展板和组合型展板两种形式。大展板须使用特殊打印机印制,缺点是成本较高,印制完毕后再修改就很麻烦,不方便储存和携带;优点是张贴方便,视觉效果良好。使用多张 A4 打印纸的组合型展板具有价格便宜、容易修改、便于携带的优点,不足之处是张贴比较费时费力。

11.3.2　会议展板的内容表达

会议展板的内容表达一般必须符合科技论文的 IMRaD 格式,通常包括题名、作者及其所属机构、摘要、引言、材料与方法、结果、结论、致谢、参考文献等部分。会议展板内容的选择要根据读者对象进行判断,如与会人员是以小同行为主还是以大同行为主,是否有潜在的合作对象等,并以此来决定所展示内容的范围、深度和侧重点。一定要避免试图在展板中塞入太多的信息,从而导致重点内容不突出、信息表达的逻辑框架缺失。

此外,在准备会议展板时,要注意插图设计和文字简洁性方面的一些问题。事实上,优秀展板的文字并不多,会议展板的大部分篇幅都用于放置图片和表格,比较好地遵循了"一图胜千言"的原则[6]。

题名:要简短、鲜明、醒目。题名的长度不可超过 2 行,应该采用粗体黑色字体,要能使读者在 3 米远的地方看清楚。

作者及所属机构:作者的次序和身份关系到贡献和知识产权,因此要确保作者的完整和所属机构的明确性。

摘要:简述研究问题、方法、结果及意义,要尽量少于 150 个单词。

引言:简介背景,提出问题,应不超过 200 个单词。

材料与方法：采用图片或流程图形式，不超过200个单词。

结果：展板的主要内容，应尽量以图表形式表达。图表要简洁、有自明性，可适当使用颜色强调。

结论：简要分段或以模型图形式表述，有时可给出进一步研究方向。为了紧凑的目的，有关讨论的内容通常糅合在结果的结论部分。

致谢：包括协助完成研究工作的机构或个人、研究资金来源等。

参考文献：采用简明形式（通常可略去论文题名）。

11.3.3 会议展板的会场讲演

会场讲演与交流不仅是展板内容的有效延伸，同时也是作者与共同兴趣同行结识和深入了解的良好渠道。

为确保展示效果，应提前准备好一个5min左右的介绍性内容，介绍的内容要结合会议展板中的图片，要强调核心问题、研究意义和重要性。要事先预想好读者可能会提出的问题及如何回答这些问题，有机会的话，也可以向读者提问，充分利用交流的机会获得同行的反馈，与同行之间建立联系。

作为展板的主人，要做到专心、开放、细致、自信，要留给参观者足够的空间和时间，因为他们要用自己的判断和节奏来阅读展板（一名仔细阅读的观众会吸引到其他人）；如果参观者提出问题，要简单、大方地予以回应，尽可能地与每位参观者保持适当的视线接触，让参观者感觉到主人的友好、亲和。

最好能准备一些与展板内容相关的详尽介绍或论文抽印本，以备分发，对展板感兴趣的读者会很感谢作者的这一做法。

11.4　国际学术会议大会发言交流

参加国际学术会议，在会议上报告研究成果并与同行进行交流，是每个年轻科学工作者所向往的事情。在国际会议上通常有两种方式报告研究成果，即会议展板（poster）及大会发言（oral presentation）。在此，暂且不谈这两种方式的差别与选择，只集中讨论在大会发言（对于新人，多数是会分在各专题报告时发言）时的方法及应注意的问题。一个人在讲演时的表现，也会体现出此人的各方面的素养，所以是要非常认真对待每次讲演[7]。

综上所述，在会议中发言最需要避免的是"我是在告诉你们这些信息（听不听得懂与我无关）"的思想，而是要树立"我想与你们分享自己的成果，希望我的报告能让你们得到或了解些什么"的观念。为了做到这一点，在讲演时（从准备讲演开始）应注意以下几个方面[8]：

（1）选择所要讲的内容的关键要点，一般会议发言只给15～30min，不可能把研究工作的方方面面都详细地讲到，所以突出重点很重要。

（2）在讲演一开始要告诉听众今天准备讲哪几方面的问题。

（3）了解听众对于要讲的问题和内容的知识程度。例如对完全对口的同行作报告时可以忽略一些专业术语的解释、对某项指标的测量方法等等,但是对仅在大专业范围一致的或者跨领域的听众作报告时,为了使大家都能听懂在说什么,必须考虑在某些地方要作深入浅出的说明。

(4)把讲演内容分成几个重点,使自己能把握好时间及节奏。

(5)要避免使用含糊和模糊不清的语言。例如避免讲:我想这个结果大概是×××意思(注意这句话中的"想""大概"均是不很肯定的意思)。自己也不确定,如何能使别人相信?

(6)在讲演中,要注意节奏与层次,使得听众能逐步深入,完全听懂讲的内容,要避免给听众过多的信息而来不及理解与接受的情况,这其实是讲演的艺术之一。准备完讲稿之后,要把自己设想为是一名听众,想一下这个讲演自己能听得懂吗?

无论每次会议有多少听众,认真准备好在科学会议上的每次讲演,是成为所在领域内一名成熟的研究者的必经之路。

11.4.1 如何准备讲演

在知道论文摘要被会议作为发言录用后,就应该立即开始着手准备,并且应是在去参加会议的一个月之前完成PPT的初稿。在剩下的一段时间内,可以再重复检查两三次讲稿,最后争取在学校里先试讲一次并听取老师和同事的意见,这种意见不仅是针对所讲的内容,而且也要对表达方式提供改进的建议。

无论会议给的讲演时间是10min还是50min,准备过程是一样的,一般有三个步骤:第一步,请工作中的好友帮忙排演一次,此次着重检查讲演的内容。着重考虑的问题:报告的内容是否全面?有没有可以省略的内容?要讲述的内容在前后次序安排上是否合适?同时要请他们其中一人帮助计时。第二步,在对上述问题做了修改以后,进行第二次排演。这次要检查:幻灯片的内容是否清楚?讲演在各节点上有否做了停顿?图片与文字,幻灯片与讲演有否很好地衔接与对应?总之,这一次主要是请同事帮助审查幻灯片方面的问题及提供改进的建议。第三步,再做一次完整的排演或试讲,此时要在讲演后安排提问和回答的环节,并在回答提问方面做些练习及准备,因为往往这是对新人来说最为困难并无从准备的一个环节。如上述,可以在学校或系里试讲时完成这一步的准备[9]。大会给定的演讲时间可以不同,但在准备过程中,必须要留出回答问题时间。

11.4.2 PPT的准备

在一般情况下,如果没有幻灯片的帮助,很难能让听众在报告时始终集中注意力,也往往难以把复杂的内容(如实验装置步骤、数据结果等)表达清楚。此外,一篇很好的讲演也可能被使用了不合适的幻灯片、讲演时遇到的设备上的技术问题等因素而搞砸。这方面的注意事项虽然是老生常谈的事情了,但是在科学会议中这方面的错误仍然会经常发生。首先要记住幻灯仅是讲演时的一个辅助手段,有的人把幻灯做得色彩缤纷、在幻灯中放入一些奇葩的标记,加上一种儿童电视中的音响效果等等,以为是别出心裁,吸引眼球,其实是在喧宾夺主,分散听众对你讲演的注意力。以下是几点建议[9]。

(1)一般规则。幻灯数量要少而精;应选择平淡整洁容易阅读的版面;在背景与文字的颜色选择方面要注意对比度的问题;一个报告的 PPT 应使用始终如一的格式:如统一的色彩、字体、布局等。

(2)文字方面的规则。一张幻灯片中以六行字最好,尽量不要超过八行,尤其会场比较大时更应注意这个问题,这样规定的目的是坐在后排的听众也能看清楚;要选择清晰的字体;避免使用大写字母(读起来比较费劲);每一行的右边要对齐,左边自由,这样做是使得每个字母的间距相等。

(3)插图方面的规则。尽量保持一个幻灯片中只有一个图;当一张图中有许多条曲线时,应该用符号、而不是用颜色来区分这些曲线;尽量要给出各点(均值)的变化大小(如方差);对坐标要加标识,说明其代表什么;避免使用三维数据图。

(4)表格方面的规则。表格要尽量简单;表格一般适宜用于发表文章但不适宜讲演;实在需要使用表格时,每列的数据或文字需对齐;对每行与每列的含义要给出说明,对测量值要注明单位。重要的数据可以凸显标出。

11.4.3　作报告时的几点建议

(1)第一张幻灯片。如同文章一样,第一张幻灯片应给出题目、作者的姓名和单位。站在发言台前,银幕上放出第一张幻灯片时,你首先要利用这个短暂的时间用目光向听众致意,然后对这一节(session)的会议主持人表示感谢,对于你的演讲题目也要说些什么。然后再换到第二张幻灯片。

(2)讲演与幻灯之间的关系。用语言去填补幻灯片中没有表达的信息,你的目的是使用语言与文字来共同表达,不需要你去读幻灯片上的文字。说到这里或许你可以明白为什么有前面说过的对幻灯片的要求。在幻灯片的文字中,你可能只需要一些关键词,但是在讲演中,你说出来的句子必须是完整的。

(3)在演讲时,需说明摘要与这个讲演的区别之处(如果有这样的情况)。许多人早早就提交了会议摘要,然后直到会议前夕又做了很多工作,或者修改了摘要中的数据及结论,因此在讲演时必须告诉听众做了哪些改动以及理由,例如可以说 Please note that these results differ from those in the printed abstract. The current results include data from an additional 50 cases。

(4)按会议规定准时结束非常重要。拖迟时间是不遵守规则的表现,这样,听众就会不耐烦地听你继续讲下去,即使你的内容再精彩也没有用了。由此可见会前演习和把握好时间的重要性。

(5)在回答问题的环节。一般会议有 5min 或 10min 的提问时间,根据你回答问题时的表现,听众们在心里会对你的知识背景有个评判。要好好把握这个环节。这个环节需说得详细一点。以下是一些建议:[9]

A. 事先要对与讲演内容有关的基础知识有充分了解与准备，否则不容易应对各种问题。

B. 最好事先能了解到听众的大致水平和情况。

C. 讲演开始时（或报告前）与会议主持人打招呼此时可能会发生作用，当有人执意坚持他的问题并与你纠缠时，可以示意主持人，请主持人帮忙解围。

D. 注意在回答问题时要避免涉及年龄、性别、宗教信仰、民族等敏感话题。

E. 当一个问题提出来以后，你可能会迅速想到提问人为什么问这个问题，此人的姿态(body language)有时会给你这种感觉。例如他是否真的需要得到答案还是为了表现他自己？但是必须记住，你的回答不是给他一个人听的，而是需要面对所有听众，所以在任何情况下，你应仅针对问题本身作答，忽略对提问者个人的看法。

F. 在出现压力时，要保持冷静。以柔克刚，不要去挑战提问者，始终与会场中的听众保持对视，以便从中获得鼓励与帮助。

G. 及时纠正提问中的错误假定，以免被提问者误导。

H. 不要将你的回答延伸到去做另外一个报告，你可以说"This topic I will present in the next meeting(or tomorrow session ××)"。

I. 不要取笑低级的问题，你可以用真诚的态度向他提供一些基础知识。

J. 如果问题很长或涉及面较大，你需要较长时间回答时，可以要求散会后私下交流。

K. 如果你知道提问者是所在领域的权威，而他的问题又较难时，你可以请他自己谈谈他的意见。

L. 如果你无法回答所提出的问题时，最好诚实地告诉听众，并且寻求是否有人可以回答这个问题。

M. 如果有人打断你的发言时，可以请会议主持人帮助，也可以说"I am coming to this particular point later."去制止那个人的纠缠，无论你是否在接下来真的会涉及此问题。

N. 如果因为会场中的设备等问题，使得不是每个人都能听清楚提问时，在你作解答前最好把问题简要地重复一遍。

（6）在国际会议作讲演需要一定外语能力，许多人可以把报告内容背出来，但是在提问环节却一点都听不懂，更不用说回答听众的提问。所以要慎重考虑是否合适用演讲的形式（不适合也可以选 poster 的方式）。参加学术会议并报告研究成果，从时间顺序上来讲，应是在完成了实验工作并初步总结了研究结果（数据处理、分析及得出初步结论）之后，为此希望去会议作交流，听取同行意见，为下一步发表文章作准备。这也是科研工作的一个环节，所以目的应该是单纯的，态度应该是真诚的。

本章参考文献

[1] 朱大明. 略论科技期刊论文的学术交流作用[J]. 中国科技期刊研究, 2006, 17(3): 481-482.

[2]何晶莹.写好科技论文 加强学术交流[J].中国电影电视技术学会影视技术文集,2007,397-401.

[3]信忠保.参加学术会议的意义[EB/OL].(2017-11-01)[2018-08-20].http://blog.sciencenet.cn/blog-58729-1083244.html.

[4]杨文志.学术会议有哪些类型和形式[EB/OL].(2009-08-26)[2018-08-20].http://blog.sciencenet.cn/home.php?do=blog&id=251432&mod=space&uid=212814.

[5]文铁峰,王新立,金红芳,等.国防科技大学国际学术交流论文统计分析[J].国防科技大学学报,1993,(1):117-120.

[6]佚名.如何制作学术会议海报[EB/OL].(2016-04-28)[2018-08-21].http://www.docin.com/p-1551373066.html.

[7]周同庆.在学术会议上作报告的一些经验[J].科技导报,2014,32(12):9.

[8]冯兆东.做好"国际学术会议报告"的几点技巧[J].科技导报,2015,33(15):112.

[9]蒋百川.在科学会议上作报告的方法简述[EB/OL].(2016-04-06)[2018-08-22].http://www.360doc.com/content/16/0406/10/6943848_548249776.shtml.

附录 A 中国各种基金项目表达方式

科学技术部

国家高技术研究发展计划(863 计划):National High-tech R&D Program of China(863 Program)

国家重点基础研究发展规划(973 计划):National Key Basic Research Program of China(973 Program)

国家 985 重点建设项目:Key Construction Program of the National "985" Project

"九五"攻关项目:National Key Technologies R & D Program of China during the 9th Five-Year Plan Period

国家基础研究计划:National Basic Research Priorities Program of China

国家科技攻关计划:National Key Technologies R & D Program of China

国家攀登计划——B 课题资助:National Climb – B Plan

国家重大科学工程二期工程基金资助:National Important Project on Science-Phase II of NSRL

教育部

国家教育部科学基金资助:Science Foundation of Ministry of Education of China

教育部科学技术研究重点(重大)项目资助:Key(Key grant)Project of Chinese Ministry of Education

国家教育部博士点基金资助项目:Ph. D. Programs Foundation of Ministry of Education of China

高等学校博士学科点专项科研基金:Research Fund for the Doctoral Program of Higher Education of China(RFDP)

国家教育部博士点专项基金资助:Doctoral Fund of Ministry of Education of China

中国博士后科学基金:Supported by China Postdoctoral Science Foundation

国家教育部回国人员科研启动基金资助:Scientific Research Foundation for Returned Scholars,Ministry of Education of China

国家教育部留学回国人员科研启动金:Scientific Research Foundation for the Returned

Overseas Chinese Scholars, State Education Ministry(SRF for ROCS, SEM)

跨世纪优秀人才计划（原）国家教委《跨世纪优秀人才计划》基金：Trans-Century Training Programme Foundation for the Talents by the State Education Commission

国家教育部优秀青年教师基金资助：Excellent Young Teachers Program of Ministry of Education of China

高等学校骨干教师资助计划：Foundation for University Key Teacher by the Ministry of Education of China

中国科学院

中国科学院基金资助：Science Foundation of the Chinese Academy of Sciences

中国科学院重点资助项目：Key Program of the Chinese Academy of Sciences

中国科学院知识创新项目：Knowledge Innovation Program of the Chinese Academy of Sciences

中国科学院"九五"重大项目：Major Programs of the Chinese Academy of Sciences during the 9th Five-Year Plan Period

中国科学院百人计划经费资助：Hundred Talents Programs of the Chinese Academy of Sciences

中国科学院院长基金特别资助：Special Foundation of President of the Chinese Academy of Sciences

中国科学院西部之光基金：West Light Foundation of The Chinese Academy of Sciences

中国科学院国际合作局重点项目资助：Supported by Bureau of International Cooperation, Chinese Academy of Sciences

中国科学院上海分院择优资助项目：Advanced Programs of Shanghai Branch, the Chinese Academy of Sciences

国家自然科学基金委员会

国家自然科学基金（面上项目；重点项目；重大项目）：National Natural Science Foundation of China(General Program; Key Program; Major Program)

国家杰出青年科学基金：National Natural Science Funds for Distinguished Young Scholar

国家自然科学基金国际合作与交流项目：Projects of International Cooperation and Exchanges NSFC

海外及香港、澳门青年学者合作研究基金：Joint Research Fund for Overseas Chinese, Hong Kong and Macao Young Scholars

其 他

国家医学科技攻关基金资助项目：National Medical Science and Technique Foundation
核工业科学基金资助：Science Foundation of Chinese Nuclear Industry
北京市自然科学基金资助：Beijing Municipal Natural Science Foundation
河南省教育厅基金资助：Foundation of He'nan Educational Committee
河南省杰出青年基金（9911）资助：Excellent Youth Foundation of He'nan Scientific Committee
黑龙江省自然科学基金资助：Natural Science Foundation of Heilongjiang Province of China
湖北省教育厅重点项目资助：Educational Commission of Hubei Province of China
江苏省科委应用基础基金资助项目：Applied Basic Research Programs of Science and Technology Commission Foundation of Jiangsu Province
山西省归国人员基金资助：Shanxi Province Foundation for Returness
山西省青年科学基金资助：Shanxi Province Science Foundation for Youths
上海市科技启明星计划资助：Shanghai Science and Technology Development Funds